Proceedings of a Conference on

Local Fields

NUFFIC Summer School
held at Driebergen (The Netherlands) in 1966

Edited by T. A. Springer

Springer-Verlag Berlin
Heidelberg GmbH
1967

Professor Dr. T. A. Springer
Mathematisch Instituut
der Rijksuniversiteit te Utrecht
Boothstraat 17, Utrecht
the Netherlands

ISBN 978-3-642-87944-9 ISBN 978-3-642-87942-5 (eBook)
DOI 10.1007/978-3-642-87942-5

Titel-No. 1434

Preface

From July 25–August 6, 1966 a Summer School on Local Fields was held in Driebergen (the Netherlands), organized by the Netherlands Universities Foundation for International Cooperation (NUFFIC) with financial support from NATO. The scientific organizing Committee consisted of F. van der Blij, A. H. M. Levelt, A. F. Monna, J. P. Murre and T. A. Springer. The Summer School was attended by approximately 80 mathematicians from various countries.

The contributions collected in the present book are all based on the talks given at the Summer School. It is hoped that the book will serve the same purpose as the Summer School: to provide an introduction to current research in Local Fields and related topics.

July 1967 T. A. Springer

Contents

Contributors

ARTIN, MICHAEL, Mathematics Department, Massachusets Institute of Technology, Cambridge, Mass., U.S.A.

BASS, HYMAN, Mathematics Department, Columbia University, New York, N.Y., U.S.A.

BRUHAT, F., 80 Bd. Pasteur, Paris xv, France

CASSELS, J. W. S., 3, Luard Close, Cambridge, England

DWORK, B., Mathematics Department, Princeton University, Princeton, N.J., U.S.A.

MAZUR, B., Mathematics Department, Harvard University, Cambridge, Mass., U.S.A.

MONNA, A. F., Mathematisch Instituut, Boothstraat 17, Utrecht, Nederland

NÉRON, A., Institut de Mathématiques, Faculté des Sciences, Orsay (S. et O.), France

RAYNAUD, M., Les Bruyères, 91, rue du Colonel Fabien, 92 Antony, France.

REMMERT, R., Mathematisches Institut der Universität, 44 Münster, Schloßplatz 2, Germany

SERRE, J.-P., 6, av. de Montespan, Paris XVI, France

SWINNERTON-DYER, H. P. F., Trinity College, Cambridge, England

TATE, JOHN T., Mathematics Department, Harvard University, Cambridge, Mass., U.S.A.

TITS, J., Mathematisches Institut der Universität, 53 Bonn, Wegelerstraße 10, Germany

VERDIER, J.-L., Institut de Mathématiques, Université de Strasbourg, Strasbourg, France.

Homotopy of Varieties in the Etale Topology[1]

M. ARTIN and B. MAZUR

This paper presents an outline of some recent work; proofs of the results we announce will be published elsewhere.

Our purpose is to study the analogues of homotopy invariants which can be obtained from varieties by using the étale topology of GROTHENDIECK. (We refer to [*1*] for the definitions and properties of this topology.) In Section 6, we give a complete description of the relation between the étale topology and the classical topology for a normal variety X over the field of complex numbers, as far as homotopy invariants are concerned. We show (6.1) that the homotopy type of X in the étale topology is a certain profinite completion (cf. Section 2) of the classical homotype type. This statement includes the classical Riemann existence theorem (cf. [*1*], XI (4.3) for instance) which asserts that every finite topological covering space of X has a unique algebraic structure (i.e., that the algebraic fundamental group is the profinite completion of the topological one), and the comparison theorem for cohomology, ([*1*], XVI (4.1)), which asserts that the classical and étale cohomologies of X with values in twisted finite coefficient groups are equal. In fact, our result is derived from these two theorems.

Two methods of obtaining homotopy information from a topology have been introduced, namely by LUBKIN [*8*], and by VERDIER [*12*] using an idea of CARTIER. Both are modifications of the Čech procedure. In this paper, we have followed Verdier's approach. However, it is clear that Lubkin's could also be used, and that the comparison theorems of Section 6 hold in his setting, in their ♮ form (cf. Section 4. The method would probably have to be modified to give the full results. On the other hand, Lubkin's method yields a more rigidified object.). LUBKIN [*8*] realized that much of the formalism of homotopy theory (eg. the Hurewicz theorem) goes through for pro-objects, but in his work the invariants are defined

[1] This research has been supported by NSF.

by passing to a limit. The basic idea of working in the pro-category itself (which is essential for our viewpoint) is due to GROTHENDIECK.

The theory we develop has applications for varieties in arbitrary characteristic of the types well known in the classical case, such as to provide obstructions to the existence of morphisms. In addition, it is related to the following question, raised by WASHNITZER: Given two varieties in characteristic 0 with isomorphic reductions modulo p, what additional conditions will imply that the varieties are themselves homotopic? Although we do not answer this question, we gain some insight into the nature of the problem (cf. Sections. 6–8).

1. Pro-objects

Recall that a *pro-object* $X=\{X_i\}$ $(i\in I)$ in a category S is a filtering inverse system of objects $X_i\in S$, i.e., a functor $I^0\to S$ where I is some filtering index category (cf. [6]). Such a pro-object X determines a functor from S to sets, written

$$\operatorname{Hom}(X, Y) \underset{\text{defn}}{=} \varinjlim_{i} \operatorname{Hom}(X_i, Y) \quad \text{for} \quad Y\in S, \tag{1.1}$$

and the pro-objects are made into a category pro-S by defining morphisms so that they correspond to morphisms of the functors (1.1). This [6] yields for two pro-objects $X=\{Y_j\}$

$$\operatorname{Hom}(X, Y) = \varprojlim_{j} \varinjlim_{i} \operatorname{Hom}(X_i, Y_j). \tag{1.2}$$

Thus pro-S is equivalent with a full subcategory of the category of functors $S\to$(Sets). Of course the representable functors are included, as is seen by taking for I the one point index category. Functors isomorphic to ones of the form (1.1) are called *pro-representable*.

It is clear that pro-S varies functorially with the category S.

We will denote by $\mathscr{H}$ the homotopy category of connected, pointed CW-complexes. This category is equivalent with the homotopy category of connected pointed Kan complexes (cf. Prop. 9.1 of [7], and [9]), and we will on occasion pass informally from one to the other. (The reader may wonder why we do not stick either to the category of CW-complexes or to the category of Kan complexes. The reason is this. On the one hand, we have occasion to make very strong use of fibre resolution and coskeleton functors, giving rise to Postnikoff decompositons etc., and these techniques have been systematically and quite elegantly developed only for Kan complexes. On the other hand, we also use the 'adjoint' techniques of attaching cells, to kill specific obstructions, etc. which have been systematically exposed only for CW-complexes.)

Let $X=\{X_i\}$ be a pro-object in $\mathscr{H}$. We define its *homotopy* and *homology* as the pro-groups and/or pro-abelian groups

$$\begin{aligned} \pi_n(X) &= \{\pi_n(X_i)\} \\ H_n(X, A) &= \{H_n(X_i, A)\} \qquad A\in(\mathrm{Ab}). \end{aligned} \tag{1.3}$$

Thus, π_n, H_n are functors on pro-$\mathscr{H}$.

To define *cohomology*, we prefer to take the limit

$$H^n(X, A) = \varinjlim_i H^n(X_i, A) \qquad A\in(\mathrm{Ab}), \tag{1.4}$$

which is an abelian group, again yielding a functor on pro-$\mathscr{H}$. Note that (1.3), (1.4) give back the old definitions for actual objects of $\mathscr{H}$.

Cohomology with values in twisted abelian groups can also be defined as follows: A homomorphism $f:\pi_1(X)\to\mathrm{Aut}(A)$ is (cf. 1.1) an element of $\varinjlim_i \mathrm{Hom}(\pi_1(X_i), \mathrm{Aut}(A))$. Thus we can define cohomology with values in the twisted coefficient group (A,f) by

$$H^n(X, A, f) = \varinjlim_{(i,\phi)} H^n(X_i, A, \phi) \tag{1.5}$$

where $\phi:\pi_1(X_i)\to\mathrm{Aut}(A)$ is a map representing f.

2. Completions

Definition (2.1): A *class* C of groups is a full sub-category of (grps) satisfying

(o) $0\in C$

(i) A subgroup of a C-group is in C. Moreover, if $0\to A'\to A\to A''\to 0$ is an exact sequence of groups, then A', $A''\in C$ if and only if $A\in C$.

We will call a class C *complete* if in addition

(ii) For A, $B\in C$, the product A^B of A with itself indexed by B is in C. (The index B is just to give a bound on the cardinality required.) We will be primarily interested in the case that C consists of finite groups, in which case (ii) follows from (i). However, we do not exclude the case that C consists of all groups.

A complete class has the following agreeable property, easily verified: If $G \supset N \supset H$ are groups such that N is normal in G and H is normal in N, and if G/N and N/H are C-groups, then H contains a subgroup K, normal in G, such that G/K is a C-group.

Let G be a group and C a class. It is immediately seen that the quotients of G which are C-groups form a pro-object which we denote by $\hat{G}$ and call the *C-completion* of G. This definition extends in an obvious way to give a functor

$$\wedge : \text{pro-(grps)} \to \text{pro-}C\,. \tag{2.2}$$

The pro-group $\hat{G}$ is characterized by the property that

$$\mathrm{Hom}(G, A) \approx \mathrm{Hom}(\hat{G}, A) \quad \text{for} \quad A \in C\,, \tag{2.3}$$

which one can express by saying that completion is *adjoint* to the inclusion of pro-C in pro-(grps).

This notion of completion has the following analogue in the homotopy category: If C is a class, let $C\mathscr{H}$ denote the full subcategory of $\mathscr{H}$ consisting of objects whose homotopy groups $\pi_q(X)$ are C-groups for each q.

Theorem (2.4): Let C be a class of groups. There is a functor

$$\wedge ; \text{pro-}\mathscr{H} \to \text{pro-}C\mathscr{H}$$

called *C-completion*, which is adjoint to the inclusion of pro-$C\mathscr{H}$ in pro-$\mathscr{H}$, i.e., such that for $X \in$ pro-$\mathscr{H}$

$$\mathrm{Hom}(X, W) \approx \mathrm{Hom}(\hat{X}, W) \quad \text{for} \quad W \in C\mathscr{H}\,.$$

For a CW-complex X, its completion $\hat{X}$ is constructed as an inverse system of CW-complexes obtained from X by spanning in enough cells in each successive dimension so that the homotopy groups of the resulting complex are all C-groups. Of course, the adjointness property gives a structure map

$$X \to \hat{X}\,.$$

This map can in fact be rigidified enough to become a pro-object in the homotopy category of *pairs* of (connected pointed) CW-complexes.

From the construction, it is clear that

$$\widehat{\pi_1(X)} \xrightarrow{\sim} \pi_1(\hat{X})\,. \tag{2.5}$$

An important fact is that for a complete class C, C-completion is compatible with "C-covering spaces" in a sense which we will not make precise here.

3. Cohomological criteria

A morphism $X \to Y$ in pro-$\mathscr{H}$ which induces an isomorphism on homotopy (cf. (1.3)) is not necessarily itself an isomorphism. To see this, let $X \in \mathscr{H}$ be a CW-complex, and denote by $X^\natural \in$ pro-$\mathscr{H}$ the inverse system

$$X^\natural = \{\mathrm{cosk}_n X\}$$

indexed by integers n, where cosk_n is the "coskeleton" functor [4] which kills homotopy in dimension $\geqq n$. We have a canonical map

$$X \to X^\natural$$

in pro-$\mathscr{H}$, and it is immediately seen from the definition (1.1) that this map is not invertible unless X has homotopy zero in high dimension. But it obviously induces an isomorphism on homotopy.

The notation $\natural$ extends in a natural way to pro-$\mathscr{H}$, It is functorial, and we will say that a map $f: X \to Y$ is a *$\natural$-isomorphism* if $f^\natural$ is an isomorphism.

Theorem (3.1): Let C be a complete class of groups, and $f: X \to Y$ a morphism in pro-$\mathscr{H}$. Then the map of C-completions $\hat{f}: \hat{X} \to \hat{Y}$ is a $\natural$-isomorphism if and only if

(i) $\widehat{\pi_1(X)} \tilde{\to} \widehat{\pi_1(Y)}$, and

(ii) for every abelian coefficient group $A \in C$ which is C-twisted on Y, the map

$$H^n(Y, A) \to H^n(X, A)$$

is bijective for all n.

Here a twisted coefficient group A is given by a map $\pi_1(Y) \to \mathrm{Aut}(A)$, and we say that A is *C-twisted* if this map factors through $\widehat{\pi_1(Y)}$.

As an application of (3.1), we get the criterion that if $X \to Y$ is a map with $Y \in$ pro-$C\mathscr{H}$, then Y is $\natural$-isomorphic to $\hat{X}$ if and only if (i) and (ii) hold. If we let C be the class of *all* groups, then completion is the identity functor and one can use (3.1) to prove

Theorem (3.2): (i) A map $f: X \to Y$ in pro-$\mathscr{H}$ is a $\natural$-isomorphism if and only if $\pi_n(f): \pi_n(X) \to \pi_n(Y)$ is an isomorphism for each n.

(ii) (Hurewicz theorem) Suppose that $\pi_q(X) = 0$ for $q < n$, $X \in$ pro-$\mathscr{H}$. Then the canonical map $\pi_n(X) \to H_n(X, Z)$ is an isomorphism.

4. Homotopy groups of completions

Let C be a *complete* class of groups, and $X \in$ pro-$\mathscr{H}$. One naturally wants to determine the homotopy of the C-completion $\hat{X}$ of X. Since

(Section 2) X maps to $\hat{X}$, there are canonical maps

$$\widehat{\pi_q(X)} \to \pi_q(\hat{X}).$$

To begin with, one has

Proposition (4.1): Suppose that $\pi_q(X)=0$ for $q<n$. Then

$$\widehat{\pi_n(X)} \approx \pi_n(\hat{X}).$$

In addition, $\pi_2(\hat{X})$ can be completely described. For simplicity, we will treat the case of a CW-complex:

Proposition (4.2): Let $X \in \mathscr{H}$. For each normal subgroup $H \subset \pi_1(X)$ such that $\pi_1(X)/H \in C$, let X_H be the covering space of X corresponding to H. Then $\pi_2(\hat{X})$ is characterized by the property that

$$\operatorname{Hom}(\pi_2(\hat{X}), A) = \varinjlim_H H^2(X_H, A) \quad \text{for all} \quad A \in C.$$

SERRE's notion ([*10*], p. 16) of good group has an obvious interpretation in the context of completions in $\mathscr{H}$:

Definition (4.3): Let G be a pro-group. G is *C-good* if for every twisted abelian $\hat{G}$-module $A \in C$, the map

$$H^q(\hat{G}, A) \to H^q(G, A)$$

is bijective for all q.

Now given a pro-group $G=\{G_i\}$, there is a canonically defined pro-object $K(G, 1)=\{L(G_i, 1)\}$ in $\mathscr{H}$ having $G=\pi_1(K)$ as its only non-vanishing homotopy. Applying (3.1), we obtain

Corollary (4.4): A pro-group G is C-good if and only if the C-completion of $K(G, 1)$ is $\natural$-isomorphic to $K(\hat{G}, 1)$.

Thus for a general progroup, the higher homotopy of $\widehat{K(G, 1)}$ measures the deviation from goodness. For example, $\pi_2(\widehat{K(\mathbb{Q}/\mathbb{Z}, 1)})=\hat{\mathbb{Z}}$, and $\pi_q=0$ for $q>2$.

It would be very interesting to obtain a detailed picture of the homotopy type of the pro-finite completions of some special groups G, for instance of arithmetic groups such as $G=SL(n, \mathbb{Z})$, whose profinite completions are well understood (cf. [2]) but which are not good in general.

To obtain results on higher homotopy, we have had to restrict C to be a class of finite groups. If $X \to Y$ is pro-object in the category of connected pointed CW-pairs, we consider the square obtained by C-completion:

$$\begin{array}{ccc} X & \to & Y \\ \downarrow & & \downarrow \\ \hat{X} & \to & \hat{Y} \end{array}$$

We show that this diagram may be rigidified in a canonical way, so as to become a pro-object in the homotopy category of squares. Thus we may "fibre-resolve" the above square in the horizontal direction to obtain pro-fibre triples,

$$\begin{array}{ccccc} F & \to & X & \to & Y \\ \downarrow & & \downarrow & & \downarrow \\ \bar{F} & \to & \hat{X} & \to & \hat{Y} \end{array}$$

Theorem (4.5): With the above notation, suppose that $Y=\{Y_i\}$ is a pro-object with each Y_i simply connected, and that C is a class of finite groups. Then the C-completion $\hat{F}$ of F is $\natural$-isomorphic to the fibre $\bar{F}$ of $\hat{X}\to\hat{Y}$.

By standard inductive procedure using Postnikov decompositions, we obtain from (4.5)

Theorem (4.6): Let C be a class of finite groups.

(i) Let $A\in$pro-(Ab). Then $K(\hat{A}, n)$ is $\natural$-isomorphic to the C-completion of $K(A, n)$ if and only if A is C-good.

(ii) Let $X\in$pro-$\mathscr{H}$ and suppose $\pi_1(X)=0$. Then the map $\widehat{\pi_q(X)}\to\pi_q(\hat{X})$ is an isomorphism for $q\leqq n$ if $\pi_q(X)$ is C-good for $q<n$.

The criterion (ii) above is false for non-simply connected spaces, even reasonable ones. For instance, let $X=S^1\amalg S^2$ be a wedge of a one-sphere and a two-sphere, and let C be the class of finite groups. The adjointness property of C-completion implies that it commutes with co-products. Hence $\hat{X}\approx\hat{S}^1\amalg\hat{S}^2$. But an analysis of the explicit formula (4.2) or direct reasoning shows that $\pi_2(\hat{X})$ is not the profinite completion of $\pi_2(X)$.

Corollary (4.7): Let $0\to A_1\to A_2\to A_3\to 0$ be an exact sequence of pro-abelian groups where A_3 is C-good for C a class of finite groups. Then $0\to\hat{A}_1\to\hat{A}_2\to\hat{A}_3\to 0$ is exact, where $\wedge$ denotes C-completion.

This corollary is obtained by applying (4.5) to the fibre triple,

$$K(A_1, 2)\to K(A_2, 2)\to K(A_3, 2).$$

5. The Verdier functor

VERDIER's definitions [*12*] are quite general. We will recall briefly his construction in the special context of the étale topology of schemes: Let X be a connected, pointed prescheme (*pointed* means that a geometric point x of X is given). Assume also that X is "locally connected for the étale topology", i.e., that for every étale [*5*] map $U\to X$, the prescheme U is a coproduct of its connected components.

Definition (5.1): A *hypercovering* $U.$ of X is a simplicial object $U.=$ $=\{U_n;\, d_i^n,\, s_i^n\}$ in the category of preschemes étale over X, satisfying the following axioms:

(i) $U_0 \to X$ is a *covering* (i.e. the map is surjective) of pointed preschemes.

(ii) For every $n \geqq 0$, the canonical map

$$U_{n+1} \to (\mathrm{cosk}_n U.)_{n+1}$$

is a covering.

It is harmless to add to these axioms also the assumption that each U_n decomposes as

$$U_n = SU_n \amalg NU_n \tag{5.2}$$

where $SU_n = (\mathrm{sk}_{n-1} U.)_n$ (sk = skeleton) is the degenerate part, and NU_n is the rest, i.e., the non-degenerate part. If each U_n is separated over X, this is in fact automatic.

Thus the "general" hypercovering is constructed as follows: Choose any $U_0 \to X$, a covering map in the category of pointed preschemes over X. We think of U_0 as being a reasonably "fine" covering. Next, extend this U_0 to a simplicial object $U_.^0$ in the category of preschemes étale over X by the coskeleton functor cosk_0. In this case, we get for U_n^0 the $(n+1)$st fibre power of U_0 over X. Now choose a (reasonably fine) covering map $NU_1 \to U_0 \times_X U_0 = U_1^0$. It suffices that NU_1 cover the complement of the diagonal of $U_0 \times_X U_0$. Put

$$U_1 = (\mathrm{sk}_0 U_.^0)_1 \amalg NU_1 .$$

This yields in a natural way a simplicial object truncated at level 1, made up of U_0, U_1. Extend it to a simplicial object $U_.^1$ by the coskeleton functor cosk_1. Next, choose a map $NU_2 \to U_2^1$ so that

$$U_2 = (\mathrm{sk}_1 U_{2}^1.) \amalg NU_2$$

covers U_2^1, and continue in this way. The desired hyper-covering $U.$ is the limit of the $U_.^n$.

By morphism of hypercoverings we mean of course just a map of simplicial objects over X preserving base points. Verdier's theorem is the following:

Theorem (5.3): Consider the category of hypercoverings of X with homotopy classes of morphisms. The opposite category of this category is filtering. Moreover, for any abelian sheaf F on X for the étale topology,

$$H^q(X, F) \approx \varinjlim_{U} H^q(F(U.)) \quad \text{for all} \quad q ,$$

where the term on the right is the cohomology of the co-simplical abelian group $F(U.)$, and where the limit is taken over all hypercovering and maps.

Actually, VERDIER [*12*] proves this theorem without reference to a base point, but the pointed case presents no novelty. An important technical point which is not explicitly stated in VERDIER, but which can be shown by the same methods, is that one can refine a hypercovering by changing it in large dimensions only:

Corollary (5.4): Let $V.$ be a fixed hypercovering. In order to obtain the cohomology $H^q(X, F)$ for $q>n$, it suffices to take the limit in (5.3) over hypercoverings $U.$ given together with a map to $V.$ and such that $U_q \xrightarrow{\sim} V_q$ for $q<n$.

The rule associating with a locally connected prescheme its set of connected components is a functor, which we denote by π_0. If we apply this functor to a hypercovering $U.$, we obtain a pointed simplicial set $\pi_0(U.)$. Now it is clear that π_0 preserves homotopies. Moreover, $\pi_0(U.)$ is connected since X is a connected prescheme. Therefore (5.2) implies that the category of hypercoverings $\{U.\}$ of the prescheme X yields under π_0 a pro-object $\{\pi_0(U.)\}$ in the homotopy category of pointed connected simplicial sets. Via the geometric realization [9], we obtain a pro-object in the homotopy category $\mathscr{H}$ of pointed connected CW-complexes, which we will denote by

$$X_{et} \in \text{pro-}\mathscr{H} \tag{5.5}$$

and will call the *homotopy type of the prescheme X for the étale topology.* We want to describe X_{et}.

6. Comparison theorems

Theorem (6.1): (Generalized Riemann existence theorem): Let X be a connected prescheme of finite type over the complex numbers, pointed by a rational point. Denote by X_{cl} (resp. X_{et}) the homotopy type of X in the classical (resp. étale) topology. There is a canonical map

$$\varepsilon: X_{cl} \to X_{et} \quad \text{in pro-}\mathscr{H}$$

and $\hat{\varepsilon}$ is an isomorphism, where $\wedge$ denotes pro-finite completion.

If X is geometrically unibranch (e.g. normal), then X_{et} is itself pro-finite, hence in this case

$$\hat{X}_{cl} \approx X_{et}.$$

From (4.6), we obtain

Corollary (6.2): In the notation of (6.1), suppose that X_{cl} is geometrically unibranch and simply connected. Then

$$\pi_q(X_{et}) \approx \widehat{\pi_q(X_{cl})}$$

for all q.

In the above theorem, the fact that $\hat{\varepsilon}$ is a $\natural$-isomorphism follows from the classical Riemann existence theorem and the comparison theorem for cohomology, using (3.1). To obtain the actual isomorphism, we give a direct argument using (5.3). A similar approach yields

Theorem (6.3): Let C be a class of groups, and let $\wedge$ denote C-completion. Let $f: X \to Y$ be a morphism of connected, locally connected, pointed preschemes. Assume that

$$\widehat{\pi_1(X)} \xrightarrow{\approx} \widehat{\pi_1(Y)},$$

and that for every C-twisted abelian coefficient group $A \in C$ on Y_{et},

$$H^q(Y_{et}, A) \xrightarrow{\approx} H^q(X_{et}, A) \quad \text{for all} \quad q \geqq 0.$$

Then

$$\hat{f}: \hat{X}_{et} \to \hat{Y}_{et}$$

is an isomorphism.

Applying the results of [1], one obtains various comparison theorems, such as

Corollary (6.4): Let X be a proper pointed scheme over $\operatorname{Spec} k$, where k is a separably algebraically closed field, and let $K \supset k$ be another separably closed field. Then the canonical map

$$\widehat{(X_K)}_{et} \to \hat{X}_{et} \quad \text{is an isomorphism},$$

where $\wedge$ denotes profinite completion.

Corollary (6.5): Let R be a discrete valuation ring with separably algebraically closed residue field k, and let $f: X \to \operatorname{Spec} R$ be a smooth proper scheme with connected geometric fibres X^1, X^0. Suppose X^1, X^0 are pointed compatibly with a chosen section of $X/\operatorname{Spec} R$. Then there is a canonical isomorphism

$$\widehat{X^1_{et}} \xrightarrow{\approx} \widehat{X^0_{et}} \quad \text{in pro-}\mathscr{H},$$

where $\wedge$ denotes completion with respect to the class of finite groups prime to the characteristic p of k.

Corollary (6.6): Let X be a connected, pointed scheme over an algebraically closed field K of characteristic zero, and let X^1, X^2 be the schemes over the complex numbers obtained from X via two embeddings of K in C. Then

$$\widehat{X^1_{cl}} \approx \widehat{X^2_{cl}}.$$

Remark: It is not necessarily true that X^1_{et} and X^0_{et} have the same homotype type, for there is a most elegant example due to SERRE [*11*] of a projective nonsingular variety V over number field K and imbeddings,

$$\varepsilon_1, \varepsilon_2 : K \to C$$

such that the induced complex projective varieties V_1, V_2 have distinct fundamental groups (and hence distinct homotopy types).

7. Reduction modulo p

Let $W \in \mathscr{H}$ be simply connected and with finite homotopy groups. Then it is not difficult to see that W can be decomposed canonically into a product over all primes p

$$W = \prod_p W_p$$

where $W_p \in \mathscr{H}$ has p-primary homotopy groups. (We think of W_p as the p-primary part of W.) It follows that if $X = \{X_i\}$ is a pro-object in $\mathscr{H}$ with each X_i simply connected, then the profinite completion $\hat{X}$ of X decomposes similarly:

$$\hat{X} = \prod_p \hat{X}_p \qquad p \in \{\text{primes}\} \tag{7.1}$$

where $\hat{X}_p$ denotes the *p-adic completion* of X, i.e., the completion with respect to the class of finite p-groups.

Now returning to the context of corollary (6.5), suppose under the hypotheses of the corollary that in addition the characteristic of the field of fractions of R is zero, and that X^1 is simply connected. The corollary implies that the l-adic completions of X^1_{et} and X^0_{et} are isomorphic for all $l \neq p$. Thus the reduction modulo the discrete valuation preserves all homotopy information prime to p.

Corollary (7.2): Let X, Y be smooth proper connected schemes over a field K of characteristic zero. Suppose given *two* discrete valuation rings R_p, R_q of K with residue fields of characteristics $p \neq q$ respectively, and that X, Y have isomorphic non-degenerate reductions modulo each of the valuations. Let X_1, Y_1 be the schemes over the complex numbers obtained from some embedding of K in C, and assume finally that X_1, Y_1 are simply connected. Then

$$(\widehat{X_1})_{cl} \approx (\widehat{Y_1})_{cl}.$$

Note that in this case we may conclude from (6.2) that the classical homotopy groups of X_1 and Y_1 are isomorphic (abstractly) since they are finitely generated abelian groups with isomorphic profinite completions.

It is clear, however, that much information of the p-adic part is lost in a reduction to characteristic p. Suppose for instance, that in the notation of (6.4) $X=\mathbb{P}_R^1$ is the projective line over Spec R. Then X_{et}^1 has the homotopy type of the profinite completion $\hat{S}^2$ of the two-sphere. But it can be shown that the etale cohomology of X^0 with p-torsion coefficients is actually *zero*. (This follows easily from ([*1*], X (5.2)). Therefore the p-adic completion of X_{et}^0 is trivial, by (6.3). Thus all p-adic information has been lost.

It is conceivable that the p-adic part could be controlled by an extension of the definition of étale homotopy, at least in the simply connected case. An interesting example of the behavior of the reduction for nonsimply connected varieties is furnished by the *Enriques*' surfaces ([*3*], Cap. VII, No. 1): These surfaces X have a unique torsion element in Pic X; it is of order 2 and yields by "Kummer theory" a covering X' of X which is principal homogeneous under the group μ_2 of square roots of unity. In characteristic $\neq 2$, this double covering X' is a generalized *Kummer surface* ([*3*], VII, No. 2), and is the universal covering space of X. Now in characteristic 2, the covering space with group μ_2 is inseparable over X. This is not so bad – the map $X' \to X$ therefore induces an étale homotopy equivalence. But the Euler characteristic $c_2(X)$ is known to be 12, whence $c_2(X')=24$ if the characteristic is not 2, whereas from the inseparable map $X' \to X$ in characteristic 2 we obtain $c_2(X')=12$. The difference in c_2 comes from the fact that, in characteristic 2, the surface X' is a Kummer surface having 12 *conical double points*, or an equivalent singularity. Thus the surface X seems to acquire "hidden singularities" in the reduction from char. 0 to char. 2, which are reflected in the change of homotopy type. The l-adic homotopy groups in characteristic 2 ($l \neq 2$), are the same as those of an Enriques' surface in characteristic zero having 6 conical double points.

8. Profinite completion and stable homotopy

Statements such as (4.6), (6.6), and (7.2) lead one to pose the following problem: Given a CW-complex, $X \in \mathscr{H}$, classify all CW-complexes X' such that the profinite completions $\hat{X}$ and $\hat{X}'$ are isomorphic. This problem can be solved in the context of *stable* homotopy, as follows: For $X = \{X_i\} \in \text{pro-}\mathscr{H}$, its *suspension* SX is defined to be the pro-object $\{SX_i\}$. The functor S has an adjoint, the loop space, from which it follows easily that C-completion commutes with suspension;

$$\widehat{SX} \approx S\hat{X}. \tag{8.1}$$

Let $X, Y \in \mathscr{H}$ be CW-complexes. The stable homotopy classes of maps from X to Y will be denoted by

$$(X, Y) = \varinjlim_{n} \operatorname{Hom}(S^n X, S^n Y). \tag{8.2}$$

Recall that (X, Y) is naturally an abelian group. We use the same definition (8.2) also when $X = \{X_i\}$ is a pro-object and Y is a CW-complex. If both $X = \{X_i\}$, $Y = \{Y_j\}$ are pro-objects, we put

$$(X, Y) = \{(X, Y_j)\} \quad \text{indexed by} \quad j, \tag{8.3}$$

where (X, Y_j) is defined by (8.2). Thus in this case (X, Y) is a *pro-abelian group*. Note that if we view X, Y as pro-objects in the stable homotopy category, then the morphisms $X \to Y$ in this category are just the elements of the group

$$\varprojlim_{j} (X, Y_j) = \text{Stable Hom}(X, Y). \tag{8.4}$$

Using (4.6) (ii) and the Puppe sequence for a cofibration, one obtains the following

Corollary (8.5): Let X, Y be finite CW-complexes, and let $\wedge$ denote profinite completion. Then (X, Y) is a finitely generated abelian group, and $(\hat{X}, \hat{Y})$ is its profinite completion.

The main result is the following:

Theorem (8.6): Let X be a finite simply connected CW-complex and denote by $E(X)$ the ring (X, X), which is a finite $\mathbb{Z}$-algebra. There is a natural $1-1$ correspondence isomorphism classes of locally free left $E(X)$-modules of rank one, and stable homotopy classes of finite CW-complexes Y such that $\hat{X}$ and $\hat{Y}$ are stably homotopic, the correspondence being given by

$$Y \rightsquigarrow (X, Y);$$

where (X, Y) is made into an $E(X)$-module by composition of maps.

It is known that the isomorphism classes of projective rank one modules over a finite $\mathbb{Z}$-algebra form a finite set [*13*]. Hence

Corollary (8.7): With the notation of (8.6), the number of classes of Y's is finite.

It can be shown by examples that the number may be greater than one.

Questions and Examples: It would be very useful to have a non-stable version of (8.6). More precisely, fixing a simply connected, finite CW-complex X, is it true that there are only a finite number of distinct homotopy types Y whose profinite completions are isomorphic to that of X? This is related to the classical fact that there are at most a finite number of integral quadratic forms of a given genus.

An interesting special case is given by studying simply connected oriented 4-manifolds M, where it is known that the oriented homotopy type of M is determined completely by the integral quadratic form $q(M)$ given by cup product on the free abelian group $H^2(M, \mathbb{Z})$. If X^1, X^2 are two such manifolds possessing isomorphic profinite completions, it follows that $q(X^1)$ and $q(X^2)$ are equivalent over $\mathbb{Z}_p$ for all primes p. Of course if X^1, X^2 are simply connected complex algebraic surfaces arising from a scheme X/k via two imbeddings $\varepsilon_1, \varepsilon_2: K \to C$, then the signatures of $q(X^1)$, $q(X^2)$ would also agree since they may be computed from the Chern classes of X. Consequently $q(X^1)$, $q(X^2)$ are of the same genus in this case. Are they equivalent over $\mathbb{Z}$?

References

[*1*] Artin, M., and A. Grothendieck: Séminaire de Géométrie algébrique 1963–64. – Cohomologie étale des Schémas. Mimeographed notes. Institute des Hautes Études Scientifiques 1963–64.

[*2*] Bass, H., M. Lazard, and J.-P. Serre: Sous-groupes d'indice fini dans $SL(n, Z)$. Bull. A.M.S. **70**, No. 3, 385–392 (1964).

[*3*] Enriques, F.: Le Superficie Algebriche. Bologna 1949.

[*4*] Gabriel, P., et M. Zisman: Séminaire Homotopique. Mimeographed notes. Université de Strasbourg 1963–64.

[*5*] Grothendieck, A.: Séminaire de Géométrie algébrique, Exposé 1. Mimeographed notes. Institute des Haute Études Scientifiques 1960–61.

[*6*] Grothendieck, A.: Technique de descente et théorèmes d'existence en géométrie algébrique. Séminaire Bourbaki, **12**, No. 195 (1959–60).

[*7*] Kan, D. M.: On Homotopy Theory and C.S.S. Groups. Annals of Math. **68**, 38–53 (1958).

[*8*] Lubkin, S.: On a Conjecture of A. Weil. American Journal of Mathematics (to appear).

[*9*] Milnor, J.: The Geometric Realization of a Semi-Simplicial Complex. Ann. of Math. **65**, 357–362 (1957).

[*10*] SERRE, J.-P.: Cohomologie Galoisienne, Lecture notes in Mathematics, No. 5, Berlin-Heidelberg-New York: Springer 1965.

[*11*] SERRE, J.-P.: Exemples de variétés projectives conjugées non homéomorphes. Comptes Rendus, Ac. Sc. Paris. **258**, 4194–4196 (1964).

[*12*] VERDIER, J.-L.: Séminaire de Géométrie algébrique. (1963–64) – Cohomologie étale des Schémas. Exposé V, appendice, Mimeographed notes, Institute des Hautes Études Scientifiques.

[*13*] ZASSENHAUS, H.: Neuer Beweis der Endlichkeit der Klassenzahl. Abh. Math. Sem. Hamburg. **12**, 276–288 (1938).

The Congruence Subgroup Problem

H. BASS

§ 1. Statement of the problem and partial solution

The results I shall describe represent work initiated by John MILNOR and myself [1], and concluded in collaboration with J.-P. SERRE [2]. SERRE's discovery that our problem is very closely related to some recent work of Calvin MOORE, attempting to generalize the "metaplectic groups" of WEIL, is presented in § 3.

We fix a global field, k, and a finite, non-empty, set S of places of k. S is assumed to contain all archimedean places, and we call S *totally imaginary* if all places in S are complex. This means k is a totally imaginary number field and that S is exactly the set of archimedean places.

A_k denotes the adèle ring of k, and A_k^S the ring of S-adèles of k; this is the restricted product of the completions k_v, $v \notin S$. We also write

$$\mathcal{O} = \mathcal{O}^S = \{x \in k \mid v(x) \geqslant 0 \quad \text{for all} \quad v \notin S\}.$$

μ_F denotes the group of roots of unity in a field F.

Let G be a simply connected simple Chevalley group, and write $\Gamma = G_{\mathcal{O}} \subset G_k$. If $\mathfrak{g}$ is a non zero ideal of $\mathcal{O}$ write

$$\Gamma_{\mathfrak{g}} = \ker(G_{\mathcal{O}} \to G_{\mathcal{O}/\mathfrak{g}})$$

and call a subgroup of Γ containing some such $\Gamma_{\mathfrak{g}}$ an *S-congruence subgroup*. These are evidently of finite index in Γ.

Congruence Subgroup Problem: *Is every subgroup of finite index in Γ an S-congruence subgroup*?

It had been expected for some time that this should have an affirmative response for G a simply connected Chevalley group of rank >1. The need to exclude rank 1 ($G = SL_2$) can be seen easily from the well known structure of the modular group, $SL_2(\mathbf{Z})/\pm 1$. Moreover the necessity of simple connectivity was demonstrated by SERRE (unpublished).

We shall describe here a complete solution of this problem for $G = SL_n$ ($n \geqslant 3$) or Sp_{2n} ($n \geqslant 2$). The proofs will appear in [2]. While the response is sometimes negative we can describe very precisely what occurs, and it seems reasonable to expect analogous phenomena for more general G.

We begin by transforming the problem. A subgroup of G_k which is commensurable with Γ will be called an *S-arithmetic subgroup* of G_k. In case k is a number field and S is the set of archimedean places then these are just the arithmetic subgroups of G_k in the sense of BOREL-HARISH-CHANDRA [3]. We obtain two Hausdorff topologies on G_k, the *S-congruence topology* and the *S-arithmetic topology*, by taking as a base for neighborhoods of 1 the S-congruence subgroups of Γ, respectively, the S-arithmetic subgroups. Since the latter refines the former there is a canonical continuous homomorphism π from the S-arithmetic completion, $\hat{G}_k$, of G_k, to the S-congruence completion. The congruence subgroup problem asks whether the two topologies coincide, or, equivalently, whether π is an isomorphism.

The S-congruence topology on G_k is clearly just the topology induced by the embedding $G_k \to G_{A_k S}$, which comes from the diagonal embedding of k in its ring of S-adeles. Since G is simply connected it follows from the strong approximation theorem of M. KNESER [4] that G_k is *dense* in $G_{A_k S}$. Consequently we can identify the S-congruence completion of G_k with $G_{A_k S}$, so π above is a homomorphism, $\pi: \hat{G}_k \to G_{A_k S}$. The closure, $\hat{\Gamma}$, of Γ in $\hat{G}_k$ is clearly just the profinite completion of Γ, so $\hat{\Gamma}$ is a compact and open subgroup of $\hat{G}_k$. Hence $\pi(\hat{\Gamma})$ is the closure of Γ in $G_{A_k S}$, which is an open subgroup. Therefore $\pi(\hat{G}_k)$ is a dense and open subgroup of $G_{A_k S}$, so π is *surjective*. Writing

$$C^S(G_k) = \ker(\pi) = \ker(\pi \mid \hat{\Gamma}),$$

therefore, $C^S(G_k)$ is a profinite group, and we have a topological extension of locally compact groups,

$$E^S(G_k): 1 \to C^S(G_k) \to \hat{G}_k \xrightarrow{\pi} G_{A_k S} \to 1.$$

Since both the right hand terms are constructed as completions of G_k the inclusion $G_k \subset G_k$ can be construed as a splitting of the extension $E^S(G_k)$ when restricted to the subgroup $G_k \subset \hat{G}_k$. Finally, we can reformulate the,

Congruence Subgroup Problem: Is $C^S(G_k) = \{1\}$?

Congruence Subgroup Conjecture: *Let G be a simply connected simple Chevalley group of rank >1. Then*

$$C^S(G_k) \cong \begin{cases} \mu_k & \text{if } S \text{ is totally imaginary} \\ \{1\} & \text{otherwise.} \end{cases}$$

Main Theorem. *The conjecture above is true for $G = SL_n$ ($n \geqslant 2$). and for $G = Sp_{2n}$ ($n \geqslant 2$).*

The proof shows also that $E^S(G_k)$ is a central extension, though this follows automatically from the fact that $C^S(G_k)$ is finite and that G_k has no finite quotients $\neq \{1\}$.

It appears rather bizarre that the totally imaginary case should behave so exceptionally, but this is given a very satisfactory explanation in § 3, where we relate this problem to the work of Calvin MOORE.

§ 2. Method of proof of the main theorem

We shall describe the method for SL_n; Sp_{2n} is handled similarly.

Let $\mathcal{O}$ be any Dedekind ring, and let $\Gamma = SL_n(\mathcal{O})$ for some $n \geqslant 3$. For an ideal $\mathfrak{g}$ of $\mathcal{O}$ set

$$\Gamma_{\mathfrak{g}} = \ker(SL_n(\mathcal{O}) \to SL_n(\mathcal{O}/\mathfrak{g})),$$

write $E_{\mathfrak{g}}$ for the normal subgroup of Γ generated by the "elementary unipotents", $I_n + te_{ij}$, $t \in \mathfrak{g}$, $i \neq j$, and set

$$C_{\mathfrak{g}} = \Gamma_{\mathfrak{g}}/E_{\mathfrak{g}}.$$

Moreover we shall think of $SL_2(\mathcal{O})$ as embedded in the upper left 2 by 2 corner of Γ. The following facts are not too difficult to prove:

$$\Gamma_{\mathfrak{g}} \text{ is generated by } E_{\mathfrak{g}} \text{ and } \Gamma_{\mathfrak{g}} \cap SL_2(\mathcal{O}). \tag{2.1}$$

$$E_{\mathfrak{g}} = [\Gamma, \Gamma_{\mathfrak{g}}]. \tag{2.2}$$

If H is a non central normal subgroup of Γ then H contains $E_{\mathfrak{g}}$ for some $\mathfrak{g} \neq 0$. (2.3)

$$\text{If } 0 \neq \mathfrak{g} \subset \mathfrak{g}' \text{ then } C_{\mathfrak{g}} \to C_{\mathfrak{g}'} \text{ is surjective.} \tag{2.4}$$

Suppose now, for a moment, that $\mathcal{O} = \mathcal{O}^S$ as in § 1. Then $\Gamma/\Gamma_{\mathfrak{g}}$ is finite, so it follows easily from (2.2) that $\Gamma/E_{\mathfrak{g}}$ is finite. Moreover (2.3) implies easily that every subgroup of finite index contains some $E_{\mathfrak{g}}$. Therefore the profinite completion of Γ is $\hat{\Gamma} = \varprojlim \Gamma/E_{\mathfrak{g}}$, while the S-congruence completion is $\bar{\Gamma} = \varprojlim \Gamma/\Gamma_{\mathfrak{g}}$. Writing $C = \ker(\hat{\Gamma} \to \bar{\Gamma})$ we have

$$C = \varprojlim C_{\mathfrak{g}},$$

and C is the group denoted $C^S(SL_n(k))$ in § 1. Finally, (2.4) implies that C maps *onto* each $C_{\mathfrak{g}}$, so the determination of C, which is the Main Theorem, is more or less equivalent to determining the $C_{\mathfrak{g}}$'s. The first theorems below give a method for doing this over any Dedekind ring. The subsequent theorems finish the calculation for $\mathcal{O}$ of arithmetic type, as above.

Again let $\mathcal{O}$ be any Dedekind ring. For an ideal $\mathfrak{g}$ of $\mathcal{O}$ set

$$W_{\mathfrak{g}} = \{(a, b) \mid (a, b) \equiv (1, 0) \bmod \mathfrak{g};\ a\mathcal{O} + b\mathcal{O} = \mathcal{O}\}.$$

There is a canonical surjection

$$\Gamma_{\mathfrak{g}} \cap SL_2(\mathcal{O}) \xrightarrow{\text{1st row}} W_{\mathfrak{g}}.$$

Let $\kappa: \Gamma_{\mathfrak{g}} \to C_{\mathfrak{g}}$ be the natural projection

Theorem 1 (Mennicke-Newman). *There is a unique map*

$$W_{\mathfrak{g}} \to C_{\mathfrak{g}}, \quad (a, b) \to \begin{bmatrix} b \\ a \end{bmatrix},$$

making

$$\begin{array}{ccc} \Gamma_{\mathfrak{g}} \cap SL_2(\mathcal{O}) & \to & \Gamma_{\mathfrak{g}} \\ {\scriptstyle \text{1st row}}\downarrow & & \downarrow{\scriptstyle \kappa} \\ W_{\mathfrak{g}} & \longrightarrow & C_{\mathfrak{g}} \end{array}$$

commutative, and it has the following properties:

MS1. $\begin{bmatrix} 0 \\ 1 \end{bmatrix} = 1$; $\begin{bmatrix} b + ta \\ a \end{bmatrix} = \begin{bmatrix} b \\ a \end{bmatrix}$ for all $t \in \mathfrak{g}$; $\begin{bmatrix} b \\ a + tb \end{bmatrix} = \begin{bmatrix} b \\ a \end{bmatrix}$

for all $t \in O$.

MS2. *If* (a, b_1), $(a, b_2) \in W_{\mathfrak{g}}$ then $\begin{bmatrix} b_1 b_2 \\ a \end{bmatrix} = \begin{bmatrix} b_1 \\ a \end{bmatrix} \begin{bmatrix} b_2 \\ a \end{bmatrix}$.

MS3. If (a_1, b), $(a_2, b) \in W$ then $\begin{bmatrix} b \\ a_1 a_2 \end{bmatrix} = \begin{bmatrix} b \\ a_1 \end{bmatrix} \begin{bmatrix} b \\ a_2 \end{bmatrix}$.

We call a function satisfying MS1, MS2, and MS3, from $W_{\mathfrak{g}}$ to a group, a *Mennicke symbol* on $W_{\mathfrak{g}}$. There is evidently a *universal* one, and the next theorem says the one above is universal.

Theorem 2. *Let $\sigma: W_{\mathfrak{g}} \to G$ be a Mennicke symbol. Then there is a unique homomorphism $f: C_{\mathfrak{g}} \to G$ such that $\sigma(a, b) = f\left[\begin{smallmatrix} b \\ a \end{smallmatrix}\right]$.*

The proof of this is rather long and tedious.

Henceforth $\mathcal{O} = \mathcal{O}^S$, as in § 1. The next theorem gives an arithmetic realization of the universal Mennicke symbol on $W_{\mathfrak{g}}$.

Theorem 3. (a) *If S is not totally imaginary then all Mennicke symbols on $W_{\mathfrak{g}}$ are trivial.*

(b) *Suppose S is totally imaginary, and let $m = [\mu_k : 1]$. Then there is an*

$r=r(\mathfrak{g})$ dividing m such that $W_{\mathfrak{g}} \to {}_r\mu_k$ (the rth roots of unity in k), defined by

$$(a, b) \to \begin{cases} \left(\frac{b}{a}\right)_r & \text{when this power residue symbol is defined,} \\ 1 & \text{otherwise,} \end{cases}$$

is a universal Mennicke symbol. If $\mathfrak{g}$ is highly divisible by m then $r(\mathfrak{g})=m$.

In case (b) we see, thanks to Theorems 1 and 2, that $C_{\mathfrak{g}} \cong {}_r\mu_K$ and $C=\varprojlim C_{\mathfrak{g}} \cong \varprojlim {}_r\mu_k \cong \mu_k$, because the $\mathfrak{g}$'s highly divisible by m are cofinal. Thus the Main Theorem is a corollary of Theorems 1, 2, and 3.

We close this section with a precise description of the integer $r(\mathfrak{g})$. Let p be a rational prime and let $n=\operatorname{ord}_p(m)$, the power to which p divides m. For $x \in \mathbb{R}$ write $[x]$ for the largest integer $\leqslant x$, and for $a \in \mathbb{Z}$ write $a_{[0,n]}$ for the nearest integer to a in the interval $[0, n]$. Then, in case (b) of Theorem 3,

$$\operatorname{ord}_p(r(\mathfrak{g})) = \min_{\mathfrak{p} \mid p} \left[\frac{\operatorname{ord}_{\mathfrak{p}}(\mathfrak{g})}{\operatorname{ord}_{\mathfrak{p}}(p)} - \frac{1}{p-1}\right]_{[0, \operatorname{ord}_p(m)]}$$

§ 3. Relationship to the work of C. MOORE

Let L be a locally compact group (we understand this to mean separable also) and let M be a locally compact L-module, i.e. a locally compact abelian group with a continuous action $L \times M \to M$. C. MOORE [5] has defined cohomology groups $H^n(L, M)$ with the usual formal properties, and the usual interpretation in low dimensions. In particular $H^2(L, M)$ classifies group extensions

$$1 \to M \xrightarrow{i} E \xrightarrow{p} L \to 1,$$

inducing the given action of L on M, where E is locally compact, p and i are continuous, and the induced maps $M \to iM$ and $E/iM \to L$ are topological isomorphisms.

An example of such an extension is

$$E^S(G_k): 1 \to C^S(G_k) \to G_k \to G_{A_k^S} \to 1$$

from § 1. Let us simplify notation by writing:

$$\bar{G} = G_{A_k^S}, \quad C = C^S(G_k),$$
$$e = (E^S(G_k)) \in H^2(\bar{G}, C).$$

The fact that $E^S(G_k)$ splits when restricted to $G_k \subset \tilde{G}$ is expressed by,

$$e \in \ker\left(H^2(\tilde{G}, C) \xrightarrow{\text{restr}} H^2(G_k, C)\right).$$

(Here we give G_k the discrete topology.) Therefore if M is a locally compact $\tilde{G}$-module and if $f \in \mathrm{Hom}_{\tilde{G}}(C, M)$, the group of continuous $\tilde{G}$-homomorphisms, then $f(e) \in \ker(H^2(\tilde{G}, M) \to H^2(G_k, M))$.

Theorem 4. *Let M be a profinite $\tilde{G}$-module. The homomorphism*

$$\mathrm{Hom}_{\tilde{G}}(C, M) \to \ker\left(H^2(G, M) \to H^2(G_k, M)\right)$$

is surjective. If G acts trivially on M and on C it is bijective.

In view of the congruence subgroup conjecture, this theorem, which is rather easy to prove, gives a complete picture of the profinite extensions of $\tilde{G} = G_{A_k^S}$ which split over G_k. Writing $k_S = \prod_{v \in S} k_v$ we have $A_k = A^S \times k_S$, so $G_{A_k} = G_{A_k^S} \times G_{k_S}$, and we can ask, similarly, for a description of profinite extensions of G_{A_k} which split over G_k. The existence of them is suggested by Weil's "metaplectic" groups [6], which are two sheeted coverings for the case $G = Sp_{2n}$. Moreover C. Moore has proved a number of interesting theorems in support of the:

Metaplectic Conjecture. *Let k be a global field, let G be a simply connected simple Chevalley group of rank >1, and let M be a profinite group on which G_{A_k} acts trivially. Then there is a natural isomorphism*

$$\mathrm{Hom}(\mu_k, M) \to \ker\left(H^2(G_{A_k}, M) \xrightarrow{\text{restr}} H^2(G_k, M)\right).$$

This amounts to saying that there is a certain central extension

$$1 \to \mu_k \to \tilde{G} \to G_{A_k} \to 1 \qquad (*)$$

which splits over G_k, and which is "universal" among all such extensions with profinite kernel. $\tilde{G}$ is then the alleged generalized metaplectic group of G.

In case k is totally imaginary with S the set of archimedean primes, then G_{k_S} is a simply connected, complex, semi-simple Lie group, so $H^2(G_{k_S}, M) = H^1(G_{k_S}, M) = 0$ for all such M as above. It follows easily that $G_{A_k} \to G = G_{A_k^S}$ induces an *isomorphism* $H^2(\tilde{G}, M) \to H^2(G_{A_k}, M)$. Thus, by virtue of Theorem 4, we see that: *for S totally imaginary, the congruence subgroup conjecture and the metaplectic conjecture are equivalent.* From this point of view the other cases of the congruence subgroup conjecture say that the 2-cocycle defining (*) depends essentially on every non-complex place v. More precisely if we restrict (*) to G_{k_v}, v non-complex, then (*) still has order $[\mu_k : 1]$ in $H^2(G_{k_v}, \mu_k)$.

References

[1] Bass, H. and J. Milnor: On the congruence subgroup problem for SL_n ($n \geqslant 3$) and Sp_{2n} ($n \geqslant 2$). Notes from Inst. for Advanced Study.

[2] Bass, H., J. Milnor, and J.-P. Serre: Solution of the congruence subgroup problem for SL_n ($n \geqslant 3$) and Sp_{2n} ($n \geqslant 2$). Publ. I.H.E.S. (to appear).

[3] Borel, A., and Harish-Chandra: Arithmetic subgroups of algebraic groups. Ann. of Math., **75**, 485–535 (1962).

[4] Kneser, M.: Strong approximation. Proceedings of Symposia in Pure Mathematics. Vol. IX. Am. Math. Soc., Providence, 187–196 (1966).

[5] Moore, C.: Extensions and low dimensional cohomology of locally compact groups, I. Trans. A.M.S., **113**, 40–63 (1964).

[6] Weil, A.: Sur certains groupes d'opérateurs unitaires. Acta Math., tome **111**, 143–211 (1964).

Groupes algébriques simples sur un corps local

F. Bruhat et J. Tits

La théorie générale des groupes algébriques semi-simples sur un corps K quelconque (racines, groupe de Weyl, sous-groupes paraboliques, BN-paire associée à un sous-groupe parabolique minimal, etc) est maintenant bien connue (cf. [*2*] et [*12*]). Notre but est d'exposer une théorie analogue lorsque K est un corps local de corps résiduel k. Un groupe algébrique simple simplement connexe G défini sur K apparait alors comme une sorte de «groupe algébrique de dimension infinie» sur le corps résiduel, plus précisément comme une limite inductive de limites projectives de variétés algébriques sur k. En particulier, on obtient dans G_K une BN-paire (B, N) de groupe de Weyl en général infini (isomorphe au groupe de Weyl affine d'un système de racines), qui est caractérisée (lorsque G n'est pas anisotrope sur K) par la propriété suivante: un sous-groupe de G_K est *borné* (au sens de la valuation de K) si et seulement si il est contenu dans la réunion d'un nombre fini de doubles classes modulo B. Ceci permet la classification (à automorphismes intérieurs près) des *sous-groupes bornés maximaux* (c'est-à-dire des sous-groupes *compacts* maximaux lorsque K est localement compact) de G_K: on trouve qu'il y en a exactement $l+1$ classes, où l est le rang relatif de G sur K.

D'autre part, notre théorie a des applications à l'étude de la cohomologie galoisienne et à la classification des formes de G sur K: elle permet en quelque sorte de ramener l'étude de la cohomologie galoisienne de G à celle de la cohomologie de groupes algébriques définis sur le corps résiduel k. En particulier, nous retrouvons et généralisons les résultats de M. Kneser relatifs aux groupes algébriques simples sur un corps p-adique [*18*].

La méthode consiste à étudier tout d'abord le cas *déployé*, où nous reprenons la théorie de N. Iwahori et H. Matsumoto [*17*], puis à passer par une «méthode de descente» d'abord au cas quasi-déployé (déjà traité par H. Hijikata [*16*]) et enfin au cas général.

1. Le cas déployé

Soit G un groupe algébrique linéaire connexe. Rappelons[1] qu'un *tore* de G est un sous-groupe algébrique T de G isomorphe à une puissance $(\text{Mult})^l$ du groupe multiplicatif Mult. Un *caractère* de T est un homomorphisme de T dans Mult. Les caractères de T forment un groupe commutatif libre de rang l, noté $X(T)$. On sait que les *tores maximaux* de G sont tous conjugués.

Nous supposerons désormais que G est *simple* (i.e. n'est pas commutatif et ne possède pas de sous-groupe distingué connexe non trivial) et *simplement connexe* (i.e. toute représentation projective de G provient d'une représentation linéaire). Soit T un tore maximal de G. Il existe un nombre fini de sous-groupes algébriques $U_1, \ldots, U_q$ de G, isomorphes au groupe additif Add et normalisés par T. Pour chaque indice j, il existe un caractère non nul a_j de T et un seul tel que, pour tout isomorphisme u_j: Add$\to U_j$, on ait:

$$t u_j(x)\, t^{-1} = u_j(a_j(t)\, x) \qquad \text{pour} \quad t \in T \quad \text{et} \quad x \in \text{Add}.$$

Ces caractères sont deux à deux distincts et leur ensemble Σ_0 forme dans l'espace vectoriel $E' = \mathbf{R} \otimes_{\mathbf{Z}} X(T)$ un *système de racines réduit irréductible* (cf. [*4*] ou [*20*]). En particulier, on a $\Sigma_0 = -\Sigma_0$, on peut choisir l éléments linéairement indépendants $a_1, \ldots, a_l$ de Σ_0 tels que tout $a \in \Sigma_0$ soit combinaison linéaire à coefficients entiers tous de même signe des a_i; un tel système est appelé une *base* de Σ_0 (ou encore un système de racines simples). Enfin, il existe une forme quadratique positive non dégénérée et une seule (à un multiple près) sur E' telle que pour tout $a \in \Sigma_0$, la réflexion orthogonale r_a par rapport à l'hyperplan orthogonal à a conserve Σ_0. Le groupe W_0 engendré par les r_a pour $a \in \Sigma_0$ est alors fini: c'est le *groupe de Weyl* de Σ_0.

Un élément $a \in \Sigma_0$ est appelé une *racine* de G suivant T et on note U_a le sous-groupe unipotent correspondant. Le normalisateur N de T opère évidemment sur Σ_0 et on montre que l'on obtient ainsi un homomorphisme de N *sur* W_0, de noyau T.

Soit maintenant E le dual de E'. Soit Σ l'ensemble des demi-espaces fermés $(a, k) = \{e \in E \mid a(e) \leqslant k\}$ pour $a \in \Sigma_0$ et $k \in \mathbf{Z}$. Un élément de Σ sera appelé une *racine affine*. Pour $\alpha = (a, k) \in \Sigma$, nous poserons $\alpha^* = (-a, -k)$, $\alpha_+ = (a, k+1)$ et nous noterons $\partial\alpha$ le bord du demi-espace α. On désigne

---

[1] Pour ces rappels sur les groupes algébriques linéaires, on pourra consulter [*1*] et [*12*].

par r_α la symétrie orthogonale (pour le produit scalaire sur l'espace affine E déduit de la forme quadratique sur E' introduite plus haut) par rapport à l'hyperplan $\partial\alpha$. Le groupe W engendré par les r_α pour $\alpha \in \Sigma$ est le *groupe de Weyl affine* de Σ_0: il opère de manière simplement transitive sur les «chambres» de E, c'est-à-dire sur les composantes connexes du complémentaire de la réunion des $\partial\alpha$. Ces chambres sont d'ailleurs des simplexes ouverts.

Supposons désormais que G est défini sur un corps K. Un tore de G est dit *déployé* sur K s'il est défini sur K et isomorphe sur K à une puissance du groupe multiplicatif. On dit que G est déployé sur K s'il possède un tore maximal T déployé sur K. Les sous-groupes U_a correspondants sont alors définis sur K et on peut choisir des isomorphismes $u_a : \mathrm{Add} \to U_a$ définis sur K et satisfaisant à des relations de commutation remarquables, qu'on trouvera dans [*11*]. Ces relations ne déterminent pas les u_a: on peut choisir arbitrairement les isomorphismes u_a pour a décrivant une base de Σ_0 et les autres sont alors déterminés au signe près. Ce choix, que nous supposons fait, correspond au choix d'une «base de Chevalley» dans l'algèbre de Lie de G, ou encore à un «épinglage» de G au sens de [*14*].

Supposons désormais que K est muni d'une valuation discrète normée v. Le groupe additif de K est alors muni d'une *filtration* à valeurs entières. Si on la transporte à U_a grâce à u_a, on obtient une filtration sur le groupe $(U_a)_K$ des points rationnels sur K de U_a. Plus précisément, nous poserons pour $(a, k) \in \Sigma$:

$$U_{(a,k)} = \{u_a(x) | x \in K, v(x) \geqslant k\}.$$

Plus généralement, à toute partie non vide Ω de E, on associe le sous-groupe U_Ω de G_K engendré par les U_α pour $\alpha \in \Sigma$ et $\alpha \supset \Omega$, et le sous-groupe P_Ω de G_K engendré par U_Ω et par le sous-groupe H formé des points $t \in T_K$ tels que $v(\chi(t)) = 0$ pour tout caractère χ de T. On voit que

(I) *pour* $\alpha, \beta \in \Sigma$ *avec* $\alpha \subsetneq \beta$, *on a* $U_\beta \subsetneq U_\alpha$; *pour* $\alpha \in \Sigma$, *on a* $\bigcap_{\beta \supset \alpha} U_\beta = \{1\}$.

D'autre part, on vérifie aussitôt que T_K/H est un groupe commutatif libre de rang l, isomorphe à $\mathrm{Hom}(X(T), \mathbf{Z})$ et l'espace vectoriel E s'identifie à $\mathbf{R} \otimes_{\mathbf{Z}} (T_K/H)$. Par suite, T_K opère par translations sur E en laissant donc invariante sa structure d'*espace affine euclidien* (qui est la seule qui sera considérée par la suite). On montre alors aisément qu'*il existe un homomorphisme* ν *et un seul de* N_K *dans le groupe des déplacements de* E, *prolongeant l'action de* T_K *sur* E, *et tel que*

(II) $nU_\alpha n^{-1} = U_{\nu(n)(\alpha)}$ *pour* $\alpha \in \Sigma$ *et* $n \in N_K$. *On a* $\nu(N_K) = W$ *et* $\mathrm{Ker}\,\nu = H$.

Par ailleurs, les relations de commutation de Chevalley [*11*] entraînent que, pour $a, b \in \Sigma_0$, $a \neq -b$, et $h, k \in \mathbf{Z}$, le groupe des commutateurs $(U_{(a,h)}, U_{(b,k)})$ est contenu dans le produit $\prod U_{(na+mb, nh+mk)}$, produit étendu aux couples (n, m) d'entiers >0 tels que $na+mb \in \Sigma_0$, rangés dans un ordre quelconque. Ceci entraîne:

(III) *Pour $\alpha, \beta \in \Sigma$, de bords non parallèles, le groupe des commutateurs (U_α, U_β) est contenu dans le groupe engendré par les U_γ pour $\gamma \in \Sigma$, $\gamma \not\supset \alpha$, $\gamma \not\supset \beta$, et $\gamma \supset \alpha \cap \beta$.*

Examinons maintenant le cas de deux racines affines $\alpha=(a, h)$ et $\beta=(b, k)$ de bords parallèles. Si $a=b$, on a $\alpha \supset \beta$ ou $\alpha \subset \beta$ et le cas a déjà été traité (cf. (I)). Si $a=-b$, on est ramené à étudier la situation dans le groupe réductif de rang semi-simple 1 engendré par U_a, U_{-a} et T, ou même dans un groupe $SL(2, K)$. On montre alors que

(IV) *Si $\beta \subsetneq \alpha^*$, on a $P_{\alpha \cap \beta} = U_\alpha H U_\beta$.*

(V) $P_{\partial\alpha} = (U_\alpha \nu^{-1}(r_\alpha) U_\alpha) \cup (U_\alpha H U_{\alpha^*+})$.

Soit C une *chambre* de E. Posons $B=P_C$ et soient $\gamma_1, \ldots, \gamma_m$ les éléments minimaux de l'ensemble des racines affines contenant C, rangés dans un ordre quelconque (si l'on prend pour C la chambre définie par l'épinglage choisi, les γ_i se composent des $(a, 0)$ pour les racines a négatives et des $(a, 1)$ pour les racines a positives). On a:

(VI) *L'application produit est une bijection de $U_{\gamma_1} \times \cdots \times U_{\gamma_m} \times H$ sur $B=P_C$.*

Enfin, comme $(U_a)_K$ est la réunion des $U_{(a,k)}$, on a:

(VII) *le groupe G_K est engendré par N_K et les U_α pour $\alpha \in \Sigma$.*

2. Données radicielles affines
Sous-groupes d'Iwahori et sous-groupes parahoriques

Soit Σ_0 un système de racines réduit et soit Σ le système de racines affines correspondant. Une *donnée radicielle affine* de type Σ dans un groupe («abstrait») Q est la donnée d'un sous-groupe N de Q, d'un homomorphisme ν de N sur le groupe de Weyl affine de Σ, de noyau noté H, et d'une famille $(U_\alpha)_{\alpha \in \Sigma}$ de sous-groupes de Q satisfaisant aux conditions (I) à (VII) ci-dessus (où l'on remplace G_K par Q et N_K par N et avec les mêmes définitions pour U_Ω et P_Ω pour $\Omega \subset E$). Nous verrons plus loin qu'à tout groupe semisimple simplement connexe G défini sur un corps local K, correspond une donnée radicielle affine bien déterminée dans G_K, obtenue par descente à partir de la donnée radicielle affine construite au n° 1 dans le cas déployé. Pour l'instant, nous allons donner quelques propriétés des données radicielles affines.

Tout d'abord, une telle donnée permet d'obtenir une BN-paire dans Q. Rappelons[2] qu'une BN-paire dans un groupe Q est la donnée de deux sous-groupes B et N de Q tels que:

(BN 1) Q est engendré par $B \cup N$;

(BN 2) $B \cap N$ est un sous-groupe distingué de N et le groupe quotient $N/(B \cap N)$ (appelé groupe de Weyl de la BN-paire) est engendré par un ensemble S d'éléments involutifs tels que:

$BrBwB \subset (BrwB) \cup (BwB)$

quels que soient $r \in S$ et $w \in N/(B \cap N)$[3];

(BN 3) $rBr \neq B$ pour tout $r \in S$.

On montre alors que les éléments $r \in S$ sont *caractérisés* par le fait que $B \cup BrB$ est un *sous-groupe*.

Théorème 1. – *Soit* $(N, \nu, (U_\alpha)_{\alpha \in \Sigma})$ *une donnée radicielle affine dans un groupe* Q *et soit* C *une chambre de* E. *Le couple* $(B = P_C, N)$ *est une BN-paire dans* Q. *On a* $H = B \cap N$ *et l'image* $\nu(S)$ *de l'ensemble* S *des générateurs involutifs distingués de* N/H *est l'ensemble des réflexions orthogonales par rapport aux faces de* C.

De là et de la théorie des BN-paires ([*22*], [*3*]), on déduit:

Corollaire 1. – *L'application* $w \longmapsto BwB$ *est une bijection de* $W = N/H$ *sur l'ensemble* $B/Q\backslash B$ *des doubles classes de* Q *modulo* B.

Corollaire 2. – *Pour tout partie* X *de* S, *désignons par* W_X *le sous-groupe de* W *engendré par les éléments* $r \in X$. *Alors* BW_XB *est un sous-groupe de* Q *et l'application* $X \longmapsto BW_X B$ *est une bijection de l'ensemble* $\mathscr{P}(S)$ *des parties de* S *sur l'ensemble des sous-groupes de* Q *contenant* B. *De plus, les sous-groupes* BW_XB *sont deux à deux non conjugués et chacun d'eux est son propre normalisateur.*

On appelle *sous-groupe d'Iwahori* de Q (relatif à la donnée radicielle affine) un conjugué de B et *sous-groupe parahorique* un conjugué d'un sous-groupe de la forme BW_XB avec W_X *fini*. Lorsque Σ_0 est irréductible, cette dernière condition équivaut à $X \neq S$ et les sous-groupes parahoriques sont alors les sous-groupes propres de Q contenant un sous-groupe d'Iwahori.

3. Immeuble de Q

Gardons les hypothèses et notations du n° précédent et supposons de

---

[2] Voir [*22*] et [*23*], ou [*3*], où une BN-paire est appelée un «système de Tits».

[3] Un élément de W est une classe modulo $B \cap N$, ce qui donne un sens aux notations BwB, etc.

plus pour simplifier que Σ_0 est *irréductible*: les chambres de E sont alors des simplexes ouverts.

On appelle *immeuble* de Q (relatif à la BN-paire (B, N)) le complexe simplicial géométrique $\mathscr{I}$ défini comme suit: les sommets de $\mathscr{I}$ sont les sous-groupes parahoriques maximaux de Q et une partie F de l'ensemble des sommets de $\mathscr{I}$ est l'ensemble des sommets d'un simplexe de $\mathscr{I}$ si et seulement si l'intersection des $P \in F$ est un sous-groupe parahorique. Les simplexes de $\mathscr{I}$ sont de dimension $\leqslant l = \dim E$ et ceux de dimension l (que nous appellerons *chambres* de $\mathscr{I}$) sont en correspondance bijective avec les sous-groupes d'Iwahori de Q. Le groupe Q opère sur $\mathscr{I}$: il suffit d'étendre par linéarité dans chaque simplexe l'action de Q par automorphismes intérieurs sur l'ensemble des sommets de $\mathscr{I}$. En utilisant le fait que chaque sous-groupe parahorique est son propre normalisateur, on voit aussitôt que le stabilisateur d'un point a de $\mathscr{I}$ est le sous-groupe parahorique intersection des sommets du simplexe (ouvert) contenant a.

On montre alors aisément qu'il existe une *injection* j de E dans $\mathscr{I}$ et une seule possédant les propriétés suivantes: l'image par j d'un sommet s de C est le sous-groupe parahorique $P_{\{s\}}$ correspondant; la restriction de j à $\bar{C}$ est une bijection affine de $\bar{C}$ sur le simplexe fermé correspondant au sous-groupe d'Iwahori P_C; pour tout $n \in N$ et tout $a \in \bar{C}$, on a $j(\nu(n)(a)) = n(j(a))$. Par définition, une *aile* de l'immeuble $\mathscr{I}$ est le transformé de l'image $j(E)$ par un élément de Q. On vérifie que si $A = g(j(E))$ est une aile de $\mathscr{I}$ (avec $g \in Q$), la structure d'espace affine euclidien et l'ensemble des demi-espaces fermés obtenus par transport de structure à partir de la structure euclidienne de E et de Σ sont indépendants du choix de g: ceci permet donc de parler des racines affines, des chambres, etc. d'une aile quelconque de $\mathscr{I}$ (on remarquera que cette définition des chambres coïncide avec celle donnée plus haut).

Théorème 2. – (i) *Deux points quelconques de $\mathscr{I}$ sont contenus dans une même aile.*

(ii) *Il existe sur $\mathscr{I}$ une distance et une seule induisant sur toute aile la distance euclidienne; elle fait de $\mathscr{I}$ un espace métrique complet.*

(iii) *Si Ω est une partie bornée non vide de l'espace métrique $\mathscr{I}$, le stabilisateur de Ω dans le groupe des isométries de $\mathscr{I}$ admet un point fixe.*

L'assertion (i) résulte facilement du corollaire 1 au théorème 1. Pour la démonstration de (ii) et (iii), ainsi que de la proposition 1 et du théorème 3 ci-dessous, voir [*6*].

Disons qu'une partie X de Q est *bornée* si elle est contenue dans une réunion finie de doubles classes modulo B. On obtient ainsi une *bornologie* (c'est-à-dire une famille $\mathscr{F}$ de parties de Q, contenant les parties finies, stable par réunions finies et telle que $X \in \mathscr{F}$ et $Y \subset X$ entraînent $Y \in \mathscr{F}$) compatible avec la structure de groupe de Q (c'est-à-dire telle que $X \in \mathscr{F}$ entraîne $X \cdot X^{-1} \in \mathscr{F}$). De plus:

Proposition 1. – *Cette bornologie est la seule qui soit compatible avec la structure de groupe de Q et pour laquelle B soit borné et Q ne le soit pas. Inversement, si (B', N') est une BN-paire dans Q dont le groupe de Weyl W' est le groupe de Weyl affine d'un système de racines réduit irréductible, alors B' et B sont conjugués.*

Les BN-paires (B, N) et (B', N') sont alors «équivalentes»: en particulier, W et W' sont isomorphes et les immeubles associés aussi (mais N et N' peuvent ne pas être conjugués).

Théorème 3. – *Soit Γ un groupe d'automorphismes de Q, conservant la classe de conjugaison des sous-groupes d'Iwahori (ou ce qui revient au même, conservant la bornologie de Q). Les conditions suivantes sont équivalentes:*

(a) *pour toute partie bornée X de Q, la réunion $\Gamma \cdot X$ des transformés de X par les éléments de Γ est bornée;*

(b) *l'ensemble $\Gamma \cdot B$ est borné;*

(c) *Γ laisse invariant un sous-groupe parahorique de Q.*

Corollaire. – *Tout sous-groupe borné de Q est contenu dans un sous-groupe borné maximal. Les sous-groupes bornés maximaux de Q sont les sous-groupes parahoriques maximaux et se répartissent en $l+1$ classes de conjugaison.*

Jusqu'à présent, notamment dans la dernière assertion du n° 2 et dans l'énoncé de la proposition 1, nous avons implicitement supposé Σ_0 non vide (cela fait d'ailleurs partie de l'hypothèse d'irréductibilité). En vue des applications du n° suivant, il est cependant indiqué d'étendre les notions introduites au cas où $l=0$ et $\Sigma_0=\emptyset$; on a alors $\Sigma=\emptyset$, $Q=B=N=H$, $W=\{1\}$, le groupe Q est borné et possède un seul sous-groupe parahorique, à savoir Q lui-même, et l'immeuble $\mathscr{I}$ est réduit à un point.

4. Descente de la donnée radicielle affine

Désormais, on désigne par K un corps *complet* pour une valuation discrète non triviale v, de corps résiduel k *parfait* et par $\tilde{K}$ l'extension non ramifiée maximale de K, de corps résiduel $\tilde{k}$.

Soit G un groupe algébrique simple, simplement connexe, défini sur K. Il existe alors une plus petite extension K' de $\tilde{K}$ telle que G soit *déployé* sur K'. Elle est galoisienne sur K; on note Γ le groupe de Galois $\mathrm{Gal}(K'/K)$.

Lemme 1. – *Il existe trois tores $S \subset S' \subset T$ de G, définis sur K, tels que S soit un tore déployé sur K maximal, S' un tore déployé sur $\tilde{K}$ maximal et T un tore maximal déployé sur K'.*

On peut alors effectuer dans $G_{K'}$ à partir du tore maximal T, les constructions du n° 1 (avec quelques modifications évidentes pour tenir compte de ce que l'extension de v à K' n'est peut-être pas normée). On obtient un système de racines affines Σ dans un espace affine euclidien E, d'espace des translations $\mathrm{Hom}(X(T), \mathbf{R})$, et une donnée radicielle affine $((\mathrm{Norm}\ T)_{K'}, v, (U_\alpha)_{\alpha\in\Sigma})$ dans $G_{K'}$, d'où une BN-paire et un immeuble $\mathcal{I}$.

On montre alors que le groupe de Galois Γ opère sur $\mathcal{I}$ en laissant fixe l'aile $j(E)$. Par transport de structure, on en déduit que Γ opère sur E en conservant la structure d'espace affine euclidien et le système Σ. Cette action de Γ sur E est compatible avec son action naturelle sur $\mathrm{Hom}(X(T), \mathbf{R})$ et on a $\sigma(U_\alpha) = U_{\sigma(\alpha)}$ pour $\sigma \in \Gamma$ et $\alpha \in \Sigma$. Le sous-espace $E^\natural$ des points fixes de Γ dans E est en particulier un espace affine sous $\mathrm{Hom}(X(S), \mathbf{R})$.

Soit $\alpha^\natural$ un demi-espace fermé de E et soit $\tilde{\alpha}^\natural$ l'intersection de toutes les racines affines $\alpha \in \Sigma$ telles que $\alpha \supset \alpha^\natural$ et $\alpha \not\supset E^\natural$. Posons:

$$U_{\alpha^\natural} = G_K \cap U_{\tilde{\alpha}^\natural}$$

(rappelons que $U_{\tilde{\alpha}^\natural}$ est le sous-groupe de $G_{K'}$ engendré par les U_α pour $\alpha \in \Sigma$ et $\alpha \supset \tilde{\alpha}^\natural$, ou encore par les U_α pour $\alpha \in \Sigma$, $\alpha \supset \alpha^\natural$ et $\alpha \not\supset E^\natural$). Définissons de plus $U^+_{\alpha^\natural}$ comme le groupe engendré par les $U_{\beta^\natural}$ pour les demi-espaces fermés $\beta^\natural$ de $E^\natural$, tels que $\beta^\natural \supsetneq \alpha^\natural$. Soit $\Sigma^\natural$ l'ensemble des demi-espaces fermés $\alpha^\natural$ de $E^\natural$ tels que $U_{\alpha^\natural} \neq U^+_{\alpha^\natural}$.

D'autre part, soit $H^\natural = \{g \in G_K | g\cdot x = x$ pour tout $x \in j(E^\natural)\}$ et $N^\natural = \{g \in G_K | g(j(E^\natural)) = j(E^\natural)\}$. On vérifie que l'action de $N^\natural$ transportée à $E^\natural$ définit un homomorphisme $v^\natural$ de $N^\natural$ dans le groupe des automorphismes de l'espace affine euclidien $E^\natural$, de noyau $H^\natural$. On pose $W^\natural = v^\natural(N^\natural)$.

Théorème 4. – (i) *Il existe un système de racines réduit $\Sigma_0^\natural$ dans l'espace vectoriel réel $X(S)\otimes\mathbf{R}$ tel que $\Sigma^\natural$ s'identifie au système de racines affines associé. Le groupe $W^\natural$ est alors le groupe de Weyl affine correspondant.*

(ii) *Le triplet $(N^\natural, v^\natural, (U_{\alpha^\natural})_{\alpha^\natural\in\Sigma^\natural})$ est une donnée radicielle affine de type $\Sigma^\natural$ dans le groupe G_K.*

(iii) *Une partie X de G_K est bornée pour la bornologie associée à cette donnée radicielle affine si et seulement si elle est bornée pour la valuation v* (i.e. *si $v \circ f$ est bornée sur X pour toute fonction f régulière sur G*).

Remarquons que le système $\Sigma_0^\natural$ n'est pas nécessairement le système des K-racines ${}_K\Sigma_0$ de G suivant S (au sens de [*2*]), mais lui est cependant relié: tout élément de $\Sigma_0^\natural$ est proportionnel à un élément de ${}_K\Sigma_0$ et réciproquement. En particulier, $W^\natural$ est extension du groupe de Weyl relatif à K de G par un groupe commutatif libre de rang l égal au K-rang de G, c'est-à-dire à la dimension de S.

On déduit alors du corollaire au théorème 3:

Corollaire. – *Tout sous-groupe borné de G_K est contenu dans un sous-groupe borné maximal et G_K possède exactement $l+1$ classes de sous-groupes bornés maximaux. En particulier, le groupe G est K-anisotrope* (i.e. $l=0$) *si et seulement si le groupe G_K est borné.*

Le théorème 4 permet de définir les *sous-groupes parahoriques* de G_K et en particulier les sous-groupes parahoriques minimaux (que nous n'appellerons pas en général sous-groupes d'Iwahori pour des raisons qui apparaîtront plus loin): ceux-ci sont deux à deux conjugués. Si L est une extension galoisienne finie de K, on peut comparer les sous-groupes parahoriques de G_K et de G_L:

Proposition 2. – *Si L est non ramifiée, les sous-groupes parahoriques de G_K sont les groupes de points rationnels sur K des sous-groupes parahoriques de G_L invariants par le groupe de Galois de L sur K. Deux sous-groupes parahoriques de G_L invariants par le groupe de Galois sont confondus (resp. conjugués dans G_L) si et seulement si leurs intersections avec G_K sont confondues (resp. conjuguées dans G_K).*

Ces assertions ne sont en général plus exactes lorsque L est une extension *ramifiée* de K.

Corollaire. – *Si L est non ramifiée et si G est K-anisotrope, G_L possède un unique sous-groupe parahorique invariant par le groupe de Galois de L sur K.*

Un sous-groupe parahorique de G_K sera appelé un *sous-groupe d'Iwahori* si pour toute extension galoisienne finie non ramifiée L de K, le sous-groupe parahorique invariant correspondant de G_L est un sous-groupe parahorique *minimal* de G_L.

5. Structures proalgébriques

On garde les hypothèses et notations du n° précédent. Supposons tout d'abord de plus que G est *déployé* sur K: on a alors $K'=\tilde{K}$ et $S=S'=T$ d'où $E=E^\natural$ etc. Soit Ω une partie bornée non vide de E:

Théorème 5. – *Il existe un schéma en groupes $\mathscr{G}_\Omega$ sur l'anneau des entiers $\mathscr{K}$ de K, affine, de type fini et lisse, et un isomorphisme φ de la fibre générique $\mathscr{G}_\Omega \otimes_{\mathscr{K}} K$ sur G tels que la bijection correspondante du groupe $\mathscr{G}_\Omega(\tilde{K})$ des points de $\mathscr{G}_\Omega$ à valeurs dans $\tilde{K}$ sur $G_{\tilde{K}}$ envoie le sous-groupe*

$\mathscr{G}_\Omega(\tilde{\mathscr{K}})$ *des points de* $\mathscr{G}_\Omega$ *à valeurs dans l'anneau des entiers* $\tilde{\mathscr{K}}$ *de* $\tilde{K}$ *sur le sous-groupe* P_Ω *de* $G_{\tilde{K}}$. *Le couple* $(\mathscr{G}_\Omega, \varphi)$ *est unique à isomorphisme près.*

Par application du «foncteur de Greenberg» (cf. [*15*]) au schéma $\mathscr{G}_\Omega \otimes_{\mathscr{K}} (\mathscr{K}/\mathfrak{p}^{n+1})$ (où $\mathfrak{p}$ désigne l'idéal maximal de $\mathscr{K}$), on obtient pour tout entier $n \geqslant 0$ un groupe algébrique noté P_Ω^n défini sur le corps résiduel k et des homomorphismes de transition $\lambda_n : P_\Omega^{n+1} \to P_\Omega^n$, définis sur k et à noyaux unipotents connexes. On démontre que les groupes P_Ω^n sont *connexes.*

Revenons maintenant au cas général. Soit L une extension galoisienne finie de K, contenue dans K' et telle que T soit déployé sur L. Le corps K' est alors l'extension non ramifiée maximale de L. Si Ω est une partie bornée non vide de $E^\natural$, donc de E, nous pouvons considérer le système projectif des groupes algébriques P_Ω^n obtenus comme ci-dessus (mais en remplaçant K par L). Le groupe de Galois Γ de K' sur K opère sur ce système projectif: plus précisément, le sous-groupe d'inertie $\Gamma_0 = \mathrm{Gal}(K'/\tilde{K})$ opère sur chacun des groupes P_Ω^n par automorphismes *algébriques*, l'intersection des images dans P_Ω^n du sous-groupe des points invariants par Γ_0 dans chacun des P_Ω^n pour $m \geqslant n$ est un sous-groupe fermé que nous noterons $(P_\Omega^\natural)^n$, enfin $\Gamma/\Gamma_0 = \mathrm{Gal}(\tilde{K}/K) = \mathrm{Gal}(\tilde{k}/k)$ opère de manière galoisienne sur chaque $(P_\Omega^\natural)^n$ et y définit une structure de *groupe algébrique défini sur k.*

Proposition 3. – *Les groupes algébriques* $(P_\Omega^\natural)^n$ *sont connexes; les applications canoniques* $\lambda_n^\natural : (P_\Omega^\natural)^{n+1} \to (P_\Omega^\natural)^n$ *sont des morphismes définis sur* k, *à noyaux unipotents connexes. Les applications canoniques de* P_Ω *dans les* P_Ω^n *définissent une bijection de* $P_\Omega \cap G_K$ *sur la limite projective des sous-groupes* $((P_\Omega^\natural)^n)_k$.

En particulier, prenons pour Ω un point de $E_\Omega^\natural$: le groupe $P = P_\Omega \cap G_K$ est alors un sous-groupe parahorique de G_K. Soit $\bar{P}$ le groupe algébrique quotient de $(P_\Omega^\natural)^0$ par son radical unipotent: c'est un groupe réductif connexe. Soit ϱ l'application canonique de P sur $\bar{P}_k$:

Proposition 4. – *L'application* $Q \mapsto \varrho^{-1}(Q_k)$ *est une bijection de l'ensemble des sous-groupes paraboliques (resp. des sous-groupes de Borel) définis sur* k *de* P *sur l'ensemble des sous-groupes parahoriques (resp. des sous-groupes d'Iwahori) de* G_K *contenus dans* P.

6. Indications sur les démonstrations des résultats précédents

Les résultats des n°s 4 et 5 ne se démontrent pas dans l'ordre où ils ont été énoncés. Comme nous l'avons déjà dit, on traite d'abord complètement le cas *déployé* (résultats du n° 1 et théorème 5). Puis on traite le cas des groupes *résiduellement déployés* sur K, i.e. des groupes G possédant un tore S déployé sur K maximal qui est aussi déployé sur $\tilde{K}$ maximal: ces groupes sont en particulier *quasi-déployés* sur K (i.e. possèdent un sous-groupe de Borel défini sur K). Sous cette hypothèse, le Lemme 1 est immédiat (on prend pour T le centralisateur de S) et on démontre le théorème 4: les conditions (IV) et (V) s'obtiennent par une étude explicite des groupes quasi-déployés de rang relatif 1, et les autres conditions du n° 1 résultent aisément de faits connus sur la structure des groupes quasi-déployés (cf. [*21*]). Ceci fait, on démontre la proposition 3, là encore par une étude explicite.

Pour passer au cas général, on remarque qu'il existe une extension galoisienne finie *non ramifiée* L de K telle que G soit résiduellement déployé sur L. Par application du théorème 3, on démontre qu'il existe un sous-groupe parahorique P de G_L qui est invariant par le groupe de Galois de L sur K. On démontre alors la proposition 3 pour le système projectif correspondant à P et on montre qu'un tore maximal de P défini sur k «se remonte» en un sous-groupe M de P dont l'image dans chaque P^n est un tore maximal. L'adhérence de Zariski de M dans G est alors un tore déployé sur $\tilde{K}$ maximal et on en déduit le Lemme 1. Ceci fait, on démontre simultanément le théorème 4 et la proposition 2 par «descente étale».

7. Diagramme de Dynkin résiduel

Supposons G résiduellement déployé sur K. On peut alors associer à G un «diagramme de Dynkin» Δ possédant les propriétés suivantes:

1) les sommets de Δ sont en correspondance bijective avec les classes de conjugaison de sous-groupes parahoriques maximaux de G_K;

2) le «diagramme de Coxeter» du groupe de Weyl $W^\natural$ s'obtient à partir de Δ en «oubliant» les flèches;

3) soit P un sous-groupe parahorique de G_K et soit $\Delta(P)$ le sous-diagramme de Δ dont les sommets correspondent à des classes de sous-groupes parahoriques maximaux ne possédant pas d'élément contenant P; alors $\Delta(P)$ est le diagramme de Dynkin (ordinaire) du groupe réductif $\bar{P}$ (qui est déployé sur le corps résiduel k).

Lorsque G est déployé sur K, le diagramme Δ n'est autre que le «diagramme de Dynkin complété» de G, i.e. le diagramme obtenu en ajoutant au diagramme de Dynkin ordinaire de G l'opposé de la plus grande racine (cf. [*4*], [*17*]).

Lorsque $l=1$, Δ n'est plus un «diagramme de Dynkin» au sens strict: il se compose de deux points unis par un trait de «multiplicité infinie». Pour $l \geq 2$, voici la liste des diagrammes de Dynkin obtenus (l'exposant en haut à gauche du type est l'ordre du groupe de Galois de la plus petite extension L de K qui déploie G: p. ex. le groupe ^2A_n est le groupe $SU(n+1)$):

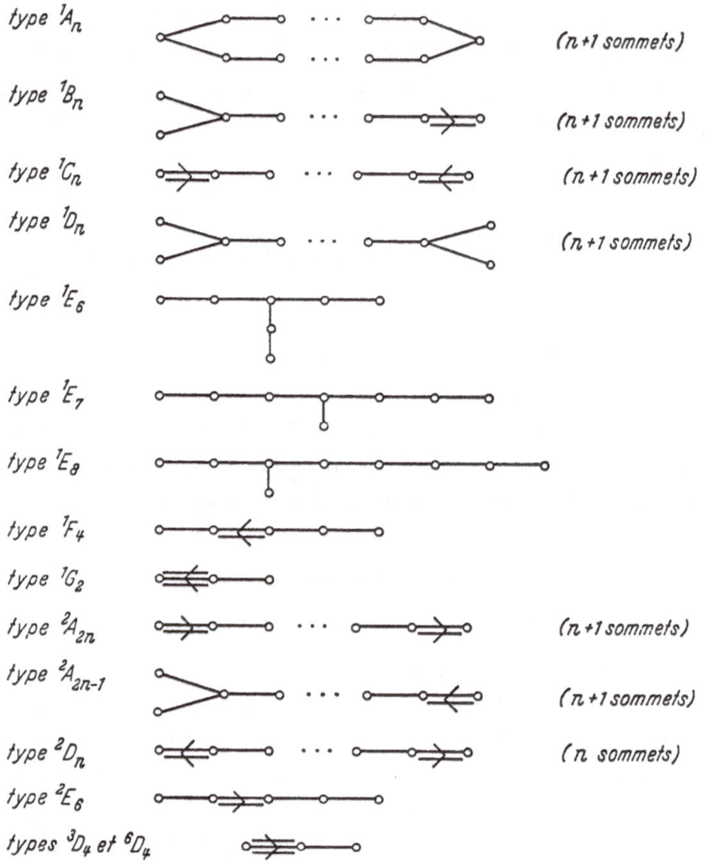

Lorsque G n'est pas résiduellement déployé, on peut lui attribuer un «indice» comme dans la théorie ordinaire des groupes semi-simples (cf. [2]): cet indice est composé du diagramme de Dynkin résiduel Δ de G considéré comme groupe défini sur une extension galoisienne L de K non ramifiée assez grande pour que G soit résiduellement déployé sur L, de l'action du groupe de Galois $\Gamma = \mathrm{Gal}(L/K)$ sur Δ traduisant son action sur l'ensemble des sous-groupes parahoriques maximaux de G_L et de l'ensemble Δ^0 des sommets de $\Delta(P)$, où P est un sous-groupe parahorique invariant minimal (dont l'intersection avec G_K est donc un sous-groupe parahorique minimal de G_K). On montre alors aisément que pour que la classe de conjugaison d'un sous-groupe parahorique Q de G_L contienne un sous-groupe invariant par Γ (donc corresponde à une classe de sous-groupes parahoriques de G_K), il faut et il suffit que $\Delta(Q)$ contienne Δ^0 et soit invariant par l'action de Γ sur Δ.

Si en particulier le groupe G est *K-anisotrope*, alors Γ *opère transitivement sur* $\Delta - \Delta^0$.

8. Applications

Proposition 5. – *Si le corps résiduel k est de dimension cohomologique* $\leqslant 1$, *le groupe* G_K *possède des sous-groupes d'Iwahori.*

Soit en effet P un sous-groupe parahorique de G_K. Comme tout groupe réductif sur un corps de dimension cohomologique $\leqslant 1$ est quasi-déployé, le groupe $\bar{P}$ possède des sous-groupes de Borel définis sur k. Il suffit alors d'appliquer la proposition 4.

Par suite, lorsque dim $k \leqslant 1$, le sous-ensemble Δ^0 de l'indice de G sur K est *vide*. Si de plus G est K-anisotrope, le groupe de Galois Γ opère donc transitivement sur le diagramme de Dynkin résiduel Δ. La liste dressée plus haut montre que cela n'est possible que si G est de type A_n et se déploie sur $\tilde{K}$. D'où:

Proposition 6. – *Si* dim$k \leqslant 1$ *et si G est K-anisotrope, alors G est de type* A_n. *Si de plus* Gal$(\tilde{K}/K)$ *est commutatif, G est une forme intérieure de* A_n *et* G_K *s'identifie au groupe multiplicatif des éléments de norme réduite* 1 *d'une algèbre à division de centre K.*

De la proposition 5, on déduit aussi, comme nous l'a signalé T. A. Springer:

Proposition 7. – *Si* dim$k \leqslant 1$, *on a* $H^1(K, G) = \{0\}$.

Lorsque K est de caractéristique zéro et k fini, les propositions 6 et 7 sont dues à M. Kneser [*18*].

Plus généralement, on peut, sans hypothèse supplémentaire sur k, ramener la détermination de $H^1(K, G)$ à celle de la cohomologie galoisienne de groupes algébriques définis sur le corps résiduel k et obtenir des résultats sur la cohomologie galoisienne de groupes non simplement connexes et la classification des formes (cf. [*10*])[4].

---

[4] On trouvera dans [*10*] d'autres applications (décompositions d'Iwasawa et de Cartan), qui sont utiles pour l'étude des représentations unitaires de G_K lorsque K est localement compact (cf. [*5*] et [*19*]).

Bibliographie

[1] Borel, A.: Linear algebraic groups. Proc. Symp. Pure. Math., vol. IX, p. 3–19. Amer. Math. Soc., Providence, R.I. (1966).

[2] Borel, A. et J. Tits: Groupes réductifs. Publ. I.H.E.S. **27**, 55–150 (1965).

[3] Bourbaki, N.: Groupes et Algèbres de Lie, chap. IV, Systèmes de Coxeter et systèmes de Tits. Paris (à paraître).

[4] Bourbaki, N.: Groupes et Algèbres de Lie, chap. VI, Systèmes de Racines. Paris (à paraître).

[5] Bruhat, F.: Sur les représentations des groupes classiques p-adiques. Amer J. Math. **83**, 321–338 et 343–368 (1961).

[6] Bruhat, F. et J. Tits: Un théorème de point fixe. I.H.E.S. Paris (1966), (miméographié).

[7] Bruhat, F. et J. Tits: BN-paires de type affine et données radicielles. C. R. Acad. Sci. **263**, 598–601 (1966).

[8] Bruhat, F. et J. Tits: Groupes simples résiduellement déployés sur un corps local. C. R. Acad. Sci. **263**, 766–768 (1966).

[9] Bruhat, F. et J. Tits: Groupes algébriques simples sur un corps local. C. R. Acad. Sci. **263**, 822–825 (1966).

[10] Bruhat, F. et J. Tits: Groupes algébriques simples sur un corps local: cohomologie galoisienne, décompositions d'Iwasawa et de Cartan. C. R. Acad. Sci. **263**, 867–869 (1966).

[11] Chevalley, C.: Sur certains groupes simples. Tohôku Math. Jour. (2), **7**, 14–66 (1955).

[12] Chevalley, C.: Séminaire sur la classification des groupes de Lie algébriques. 2 vol., Paris, Inst. H. Poincaré, 1958 (miméographié).

[13] Chevalley, C.: Certains schémas de groupes semi-simples, Séminaire N. Bourbaki, **13**, exposé 219 (1960–61). New York: Benjamin 1966.

[14] Demazure, M. et A. Grothendieck: Schémas en groupes. I.H.E.S. 1964 (miméographié).

[15] Greenberg, M.: Schemata over local rings. Annals of Math. **73**, 624–648 (1961); **78**, 256–266 (1963).

[16] Hijikata, H.: On the arithmetic of p-adic Steinberg groups. Yale Univ. (1964), (miméographié).

[17] Iwahori, N. et H. Matsumoto: On some Bruhat decomposition and the structure of the Hecke ring of p-adic Chevalley groups. Publ. I.H.E.S. **25**, 5–48 (1965).

[18] Kneser, M.: Galois-Kohomologie halbeinfacher Gruppen über p-adische Körpern. Math. Zeit. **89**, 250–272 (1965).

[19] Satake, I.: Theory of spherical functions on reductive algebraic groups over p-adic fields. Publ. I.H.E.S., **18**, 5–69 (1963).

[20] Serre, J. P.: Algèbres de Lie semi-simples complexes. New York: Benjamin 1966.

[21] Steinberg, R.: Variations on a theme of Chevalley. Pacific J. Math. **9**, 875–891 (1959).

[22] Tits, J.: Théorème de Bruhat et sous-groupes paraboliques. C.R. Acad. Sci. **254**, 3419–3422 (1962).

[23] Tits, J.: Algebraic and abstract simple groups. Annals of Math., 80, 313–329 (1964).

[24] Tits, J.: Structures et groupes de Weyl. Séminaire N. Bourbaki, **17**, exposé 288 (1964–65). New York: Benjamin 1966.

Elliptic Curves Over Local Fields

J. W. S. Cassels

Introduction

We give only a brief resumé of what was said at the Summer School and refer to the author's long survey article [1] for further details and for all references to the literature.

1. Geometry

An elliptic curve $\mathscr{C}$ defined over a field k and with a point $\mathfrak{o}$ defined over k (a "rational point") has a natural structure as an algebraic group defined over k, the point $\mathfrak{o}$ playing the part of zero. This group-law is defined by writing $\mathfrak{a}+\mathfrak{b}=\mathfrak{c}$ when the divisors $(\mathfrak{a})(\mathfrak{b})$ and $(\mathfrak{c})(\mathfrak{o})$ are linearly equivalent. With this structure $\mathscr{C}$ is an abelian variety[1] of dimension 1, and all abelian varieties of dimension 1 are of this kind.

When char $k \neq 2, 3$ there is a birational equivalence over k of $\mathscr{C}$ with a cubic curve

$$y^2 z = x^3 + Axz^2 + Bz^3 \qquad (A, B \in k), \tag{1}$$

the given rational point $\mathfrak{o}$ going into the point $(x,y,z)=(0, 1, 0)$, that is, the unique point of the curve at infinity in the inhomogeneous coordinates with $z=1$. We have now $\mathfrak{a}+\mathfrak{b}+\mathfrak{c}=\mathfrak{o}$ on (1) if and only if the three points $\mathfrak{a}$, $\mathfrak{b}$, $\mathfrak{c}$ are collinear.

Any elliptic curve $\mathscr{D}$ defined over k not necessarily with a rational point belongs to a unique abelian variety $\mathscr{C}$ also defined over k, namely its jacobian. The notion of jacobian in this context is rather more refined than the classical one since it takes into account fields of definition. Fur-

---

[1] More precisely $\mathscr{C}$ is an abelian variety if it is a nonsingular curve. But in this survey we shall be concerned with curves only up to birational equivalence, so will retain the term "abelian variety" and suppress the "of dimension 1" since we are not interested in any other dimension.

ther, $\mathscr{D}$ has at least one and at most six structures as principal homogeneous space over $\mathscr{C}$, that is there is a rational map

$$\mu: \mathscr{D} \times \mathscr{C} \to \mathscr{C},$$

say

$$(\mathfrak{y}, \mathfrak{x}) \to \mu(\mathfrak{y}, \mathfrak{x}),$$

which satisfies

$$\mu(\mathfrak{y}, \mathfrak{o}) = \mathfrak{y}$$

$$\mu(\mu(\mathfrak{y}, \mathfrak{x}_1), \mathfrak{x}_2) = (\mathfrak{y}, \mathfrak{x}_1 + \mathfrak{x}_2)$$

and which has an inverse rational map

$$\nu: \mathscr{D} \times \mathscr{D} \to \mathscr{C}$$

such that

$$\mu(\mathfrak{y}_1, \mathfrak{x}) = \mathfrak{y}_2$$

is the same as

$$\nu(\mathfrak{y}_2, \mathfrak{y}_1) = \mathfrak{x}.$$

Here μ and ν are both defined over k.

When k is perfect, there is a bijection of the set $\mathrm{WC}(\mathscr{C}, k)$ of principal homogeneous spaces with given $\mathscr{C}$ and k onto the cohomology group

$$H^1(\Gamma, \mathfrak{G}), \tag{2}$$

where the symbols have the following meanings:

Γ is the galois group of $\bar{k}/k$ where $\bar{k}$ is the algebraic closure of k.

$\mathfrak{G}$ is the group of points of $\mathscr{C}$ defined over $\bar{k}$.

H^1 is the groups of continuous cocycles modulo coboundaries when Γ is given the usual (Krull) topology and $\mathfrak{G}$ the *discrete* topology.

If $\mathscr{D}$ is an elliptic curve with the structure μ of homogeneous space, the corresponding element of (2) is given by the cocycle

$$\mathfrak{a}_\sigma = \nu(\mathfrak{A}, \sigma\mathfrak{A})$$

where $\mathfrak{A}$ is any point of $\mathscr{D}$ defined over $\bar{k}$ and σ runs through Γ.

We endow $\mathrm{WC}(\mathscr{C}, k)$ with the natural group structure of (2). It is thus an abelian torsion group. Further, if K is any field containing k there is a natural map

$$j_{K/k}: \mathrm{WC}(\mathscr{C}, k) \to \mathrm{WC}(\mathscr{C}, K), \tag{3}$$

since anything which is defined over k is also defined over K. It turns out that (3) is a group homomorphism but in general is neither an injection nor a surjection.

2. Over local fields

When k is $\mathbb{C}$, the field of complex numbers, the structure of the group

$\mathfrak{G}$ of rational points is classical. The Weierstrass elliptic functions give a parametrization and so $\mathfrak{G}$ is isomorphic to $\mathbb{C}/\wedge$, where $\wedge$ is a lattice depending on $\mathscr{C}$.

The elliptic functions also give a parametrization of $\mathfrak{G}$ when k is the field $\mathbb{R}$ of reals. In this case $\mathfrak{G}$ is isomorphic to $\mathbb{R}/\mathbb{Z}$ if $\mathscr{C}$ has one real branch and to $(\mathbb{R}/\mathbb{Z})\oplus(\mathbb{Z}/2\mathbb{Z})$ if it has two.

For a non-archimedean local field there is a parametrization formally identical with the Weierstrass parametrization which is, however, valid only in a neighbourhood of $\mathfrak{o}$. It turns out that $\mathfrak{G}$ has a subgroup $\mathfrak{U}$ of finite index which is isomorphic to the additive group of integers of k. This implies, in particular, that the torsion subgroup of $\mathfrak{G}$ is finite.

In every case $\mathfrak{G}$ is a compact group in the topology inherited from k, and it is totally disconnected in the non-archimedean case.

The group WC (with discret topology) is canonically isomorphic to the group of continuous characters of $\mathfrak{G}/\mathfrak{B}$, where $\mathfrak{B}$ is the connected component of the identity (so consisting only of the zero element except for $\mathbb{C}$ and $\mathbb{R}$).

3. Relationship to global fields

When k is a global field the group $\mathfrak{G}$ has long been known to be finitely generated but there is no effective procedure known for finding the number of generators.

It is natural to consider the behaviour of the WC group under the maps (3) where K runs through all the local completions of k. An important part is played by the Tate-Šafarevič group Ш, the intersection of the kernels of (3) when K runs through all local completions. The elements of Ш correspond to homogeneous spaces $(\mathscr{D}, \mu)$, for which there is a point on the curve $\mathscr{D}$ defined over each local completion. It is known that Ш may consist of more than one element [i.e. the Hasse Principle fails] but it is conjectured that it is always finite. The structure of Ш is closely related to the problem of finding the group $\mathfrak{G}$ effectively.

For the many interesting conjectures and the few theorems in this circle of ideas see [*1*] and Swinnerton-Dyer's contribution to this volume.

Reference

[*1*] Cassels, J. W. S.: Diophantine equations with special reference to elliptic curves. J. London Math. Soc. **41**, 193–291 (1966).

On the Rationality of Zeta Functions and L-Series

B. DWORK

Let V be an algebraic variety, defined over $GF[q]$. We recall the definition of the zeta function of V,

$$Z(V,t) = \exp\left(\sum_{s=1}^{\infty} N_s t^s/s\right)$$

where N_s is the number of points of V which are rational over $GF[q^s]$. (For definition of L-series, see chapter II). It has been known for some time [*1*] that the zeta function is rational and a second exposition of the same proof has been given by SERRE [*2*]. It is rather questionable as to whether there is any need to repeat such well known material. However because of its connection with p-adic analysis it may be in accord with the purpose of this conference to outline the old proof. This will be done in chapter I but we will use results and points of view which were not available in 1959.

The members of this conference are of course aware of the independent treatment [*10*] of this problem by A. GROTHENDIECK, M. ARTIN and J. VERDIER which has been reported seperately at this conference. We hope that the present article may eliminate some misconceptions which may arise from [*10*].

It was noted by SERRE that the methods of [*1*] could be immediately applied to certain L-series of galois coverings of a variety defined over $GF[q]$. Here however the relation between the galois group, G, of the covering and the field of definition is critical.

Theorem 1 (SERRE). Let M be the set of all prime powers, m, which occur as an order of a cyclic factor group of a subgroup of G. Suppose that for each $m \in M$, the polynomial, $t^m - 1$, splits in $GF[q]$. Then each L-series of the covering is the product of a rational function with the L-series of coverings of varieties of lower dimension.

Thus, in particular, it was known in 1960 that for any galois covering, the L-series are rational provided the field of definition of the varieties is sufficiently enlarged. This proof of Theorem 1 will be sketched in chapter II.

By generalizing our trace formula, REICH [3], extended the above theorem to give rationality under the weaker hypothesis:

(R) For each $m \in M$ there exists an integer, l, prime to m such that $t^m - 1$ splits in $GF[q^l]$.

The work of REICH will be discussed in chapter III.

In chapter IV we sketch a proof of the rationality of all L-series based upon the indications in our Woods Hole talk (1964) of how our deformation theory (when applied to families of polynomials in one variable) may be used to investigate L-series. For purposes of exposition we devote particular attention to those cyclic coverings of affine n-space in characteristic p which may be lifted to cyclic coverings of affine n-space in characteristic zero. For these cases the Reich trace formula is adequate, but to complete the proof the method must be extended to cyclic coverings of hypersurfaces. For this extension (§ 5) we are fortunate in being able to appeal to a special case of a general trace formula recently reported by MONSKY.

Throughout this report we use induction on dimension and reduce the question of rationality of L-series to that of L-series of varieties of lower dimension. For the same reason no distinction is made between the geometric and algebraic function field point of view.

I. Zeta Function

All proofs of rationality of zeta functions may be said to be based on the idea of showing that the zeros and poles of a zeta function lie among the eigenvalues of a finite set of endomorphisms. Rationality would not ordinarily follow unless it were further known that the endomorphisms have finite rank. The simplicity of the p-adic method (as compared to the so called "cohomological method") lies in the fact that it is enough if the endomorphisms are completely continuous (i.e. limits in a suitable sense of endomorphisms of finite rank). On the other hand this simplification is directly useful only for the proof of rationality. For the verification of the remaining conjectures of Weil it is necessary to construct appropriate endomorphisms of finite rank. This construction has been carried out in a number of cases and it is by this time clear that the p-adic method is certainly cohomological, that is there exists an underlying p-adic cohomological theory. This aspect will not be treated in the present report.

The simplification referred to above is accomplished by means of a generalization of a theorem of E. BOREL (1894) which asserts that a power series with coefficients in the ring of integers of an algebraic number field certainly represents a rational function if it has non-zero radius of convergence at each archimedian valuation and if for at least one finite prime p it is meromorphic everywhere (i.e. is the quotient of two power series which converge p-adically for all finite values of the variable). This follows from the method of Borel: Rationality is implied by the vanishing of certain determinants formed from the coefficients of the given power series. Borel noted that in estimating the magnitude of these determinants one may use the radius of meromorphy (instead of the radius of convergence, which gives weaker estimates). This is also valid for p-adic estimates and the idea of the proof is to show that under the stated conditions, the determinants are elements of an algebraic number field (obviously) which fail to satisfy the product formula and hence must be zero.

In the following we need therefore only consider the problem of showing that L-series and zeta functions are p-adically meromorphic where p is the characteristic of the field of definition of the varieties in question. The connection between these functions and completely continuous endomorphisms of p-adic Banach spaces is important because (using SERRE [4]) the Fredholm determinants of such endomorphisms are p-adically entire.

The representation of zeta functions by means of endomorphisms requires a trace formula. In his proof that finite fields are C_1, WARNING [5] used the fact that if $\bar{f}$ is an element of $(GF[q])[X_1, \ldots, X_n]$ then N_1', the number of rational points (in affine n-space) not on the hypersurface

$$\bar{f}(X) = 0,$$

satisfies the relation

$$N_1' = \Sigma \bar{f}(x)^{q-1} \qquad \bmod p,$$

the sum being over all rational points in the ambient space.

This can be improved:

$$N_1' \equiv \Sigma f(x)^{(q-1)p^r} \quad \bmod p^{r+1} \tag{1}$$

where f is a lifting of $\bar{f}$ to a polynomial in an absolutely unramified field whose residue class field is $GF[q]$ and the sum is over all $x = (x_1, \ldots, x_n)$ such that $x_i^q = x_i$. This observation led [6] to an elementary trace formula which we now describe.

Let $\mathfrak{F}$ be a space of functions, let ψ be a mapping of $\mathfrak{F}$ (perhaps into some other space of functions) defined by

$$(\psi\xi)(X) = q^{-n}\Sigma\xi(Y),$$

the sum being over all Y such that $Y^q = X$. (We refrain from being more explicit as the mapping will be used in a number of situations). If $\mathfrak{F}$ is the space of polynomials in n variables over a field of characteristic zero and if G is a fixed element of $\mathfrak{F}$ then both ψ and $\psi \circ G((\psi \circ G)\,\xi = \psi(G\xi))$ are endomorphisms of $\mathfrak{F}$ with a well defined trace and

$$(q-1)^n \operatorname{Tr}(\psi \circ G) = \Sigma G(x), \tag{2}$$

the sum being over all $(x_1, \ldots, x_n)$ such that $x_i^{q-1} = 1$ of each i. Thus

$$(q-1)^n \operatorname{Tr}(\psi \circ f^{(q-1)p^r}) \equiv N_1'' \qquad \operatorname{mod} p^{r+1} \tag{3}$$

where N_1'' is the number of rational points in the ambient space with no zero coordinate and which do not lie on the hypersurface.

Thus N_1'' is a limit of traces but because of a failure to estimate the rapidity of convergence of (3) as $r \to \infty$, the investigation, [*6*], did not lead to a proof of meromorphy. We shall return to this in discussing the work of Reich.

The trace formula (2) can be generalized. Let Ω be the completion of the algebraic closure of the p-adic rationals, let $b > 1$ be a real number, let

$$D(b) = \{x \in \Omega^n \mid |x_i| \leqslant b,\ i = 1, 2, \ldots n\}$$

and let $\mathfrak{F}(b)$ be the space of functions holomorphic on $D(b)$. Then ψ gives a mapping of $\mathfrak{F}(b^{1/q})$ into $\mathfrak{F}(b)$ and if $G \in \mathfrak{F}(b^{1/q})$ then $\psi \circ G$ is a completely continuous endomorphism of $\mathfrak{F}(b)$, its trace may be defined in terms of the corresponding infinite matrix giving the action of $\psi \circ G$ on monomials, equation (2) remains valid and $\det(I - t(\psi \circ G))$ (which we may define to be $\exp\{\sum_{s=1}^{\infty} t^s \operatorname{Tr}(\psi \circ G)^s/s\}$) is entire. The factor $(q-1)^n$ appearing in (2) causes no difficulty and hence to prove rationality of zeta functions it is enough to be able to write N_1, the number of rational points of a hypersurface, in terms of a sum such as appears on the right side of (2) where $G \in \mathfrak{F}(b)$ for some $b > 1$.

To obtain such an expression for N_1, we may appeal to the conventional method involving sums of additive characters,

$$qN_1 = \Sigma\theta_1(\bar{x}_0 f(\bar{x})) \tag{4}$$

where θ_1 is a non-trivial additive character of $GF[q]$ and the sum on the right hand side is over all $\bar{x}_0, \bar{x}_1, \ldots, \bar{x}_n$ in $GF[q]$. The right side of (4) is converted into a sum of the required type by showing that if π is chosen with $\pi^{p-1}=-p$ then

$$\theta(t)=\exp\pi(t-t^q) \tag{5}$$

is a power series in one variable converging in a disk of radius greater than unity and that if for each $\bar{t}$ in $GF[q]$, $T(\bar{t})$ denotes the unique root (in a field of characteristic zero) of $t^q-t=0$ which lies in the residue class, $\bar{t}$, then

$$\bar{t}\to\theta(T(\bar{t})) \tag{6}$$

is a non-trivial additive character of $GF[q]$.

II. Theorem of SERRE

Let $\bar{K}=GF[q]$, V be a variety defined over $\bar{K}$, $\bar{F}$ the field of algebraic functions of V and $\bar{E}$ a galois extension of $\bar{F}$. The relation of being conjugate over $\bar{K}$ gives a partition of the algebraic points of V into primes. The degree of a prime is defined to be the number of elements in the class and each algebraic point, x, of V is said to have a degree, $\deg x$, equal to the degree, $\deg \mathfrak{p}_x$, of the class, $\mathfrak{p}_x$, in which it lies. If $\mathfrak{p}$ is a prime of V which does not ramify in $\bar{E}$ then we may associate with $\mathfrak{p}$, the class, $\left(\frac{\bar{E}/\bar{F}}{\mathfrak{p}}\right)$, of conjugate elements of the galois group $G(\bar{E}/\bar{F})$ and if χ is a character of the galois group then for each integer r, we may define $\chi(\mathfrak{p}^r)$ uniquely as being the value assumed by χ at the rth power of any element of $\left(\frac{\bar{E}/\bar{F}}{\mathfrak{p}}\right)$. The L-series may be defined

$$L(s,\chi,\bar{E}/\bar{F})=\exp\{\sum_{s=1}^{\infty} t^s/s\,\Sigma\chi(p_x^{s/\deg x})\} \tag{7}$$

the inner sum being over all $x\in V$ such that $\mathfrak{p}_x$ does not ramify in $\bar{E}$ and such that $\deg x$ divides s.

We recall that there is a better definition of L-series which also contains the contribution of ramified prime, but for proofs of rationality this may be disregarded by means of a suitable induction hypothesis. The question of rationality can be reduced to the case in which $\bar{E}$ is a cyclic extension of $\bar{F}$ by use of the linearity, factor group, induced character properties of L and by use of Brauers' decomposition of characters into Z-linear combinations of characters induced by linear characters of subgroups. To obtain by the methods of Chapter I the result of SERRE (theorem 1 above) we need a further reduction to the case in which $\bar{E}/\bar{F}$ is cyclic of prime power degree. This reduction does not follow from general principles but rather from the technique itself. We ignore this technical point and instead briefly explain the case in which $\bar{E}$ is a Kummer extension of $\bar{F}$ of degree m, prime to p.

In this Kummer case, $\bar{E}=\bar{F}(\alpha)$, $\alpha^m=\bar{f}\varepsilon\bar{F}$ and we may suppose that V is imbedded in affine space and that $\bar{f}$ is the restriction to V of a polynomial, so that $\bar{f}$ is defined everywhere on V. If x is a point of V at which $\bar{f}$ does not vanish and if x is rational over $\bar{K}$, then by an elementary computation.

$$\alpha^{-1}\left(\frac{\bar{E}/\bar{F}}{\mathfrak{p}_x}\right)\alpha=\bar{f}(x)^{(q-1)/m} \tag{8}$$

Thus using T once again to denote Teichmüller representative, the computation is reduced to that of sums of the form

$$\Sigma T\left(\bar{f}(x)^{(q-1)/m}\right) \tag{9}$$

where x runs over rational points of V. Now with θ_1 as in (4)

$$qT\left(\bar{f}(x)^{(q-1)/m}\right)=\Sigma(Ty)^{(q-1)/m}\,\theta_1\left(z\left(y-\bar{f}(x)\right)\right), \tag{10}$$

the sum on the right being over all y and z in $\bar{K}$. This may be put in analytical form by means of (5) and the condition $x\varepsilon V$ may be expressed analytically by means of the method of Chapter I, the final result being again a sum such as in the right side of (2) with G in $F(b)$ for some $b>1$, but here G is a function of Y and Z as well as the other variables.

III. Trace formula of Reich

As before, let $\bar{K}=GF[q]$, let K be the unramified extension of the p-adic rationals whose residue class field is $\bar{K}$. Let $f(X_1,\ldots,X_n)$ be a polynomial of degree d with coefficients in the ring of integers of K. Let ε, Δ be non-negative rational numbers such that

$$1>\varepsilon+d\Delta$$

Put

$$D_{\varepsilon,\Delta,f}=\{x=(x_1,x_2,\ldots,x_n)|\ \operatorname{ord}f(x)\leqslant\varepsilon,\quad \operatorname{ord}x_i\geqslant-\Delta,\ i=1,2,\ldots,n\} \tag{11}$$

and let $F_{\varepsilon,\Delta,f}$ be the space of all functions holomorphic on $D_{\varepsilon,\Delta,f}$ (i.e.

uniform limits of rational functions on $D_{\varepsilon, \Delta, f}$). Then ψ defines a continuous map of $F_{\varepsilon/q, \Delta/q, f}$ into $F_{\varepsilon, \Delta, f}$ and if

$$G \in F_{\varepsilon/q, \Delta/q, f} \quad \text{then} \quad \psi \circ G : \xi \to \psi(G\xi)$$

is a continuous endomorphism of $F_{\varepsilon, \Delta, f}$. For the applications we may suppose that f is homogeneous and that the reduced polynomial $\bar{f}$ splits into distinct irreducible factors in the algebraic closure of $\bar{K}$. With this hypothesis, the results of Reich state

Theorem 2. (a) If $\varepsilon > 0$, $\Delta > 0$, then $\psi \circ G$ is a completely continuous endomorphism of $F_{\varepsilon, \Delta, f}$

(b) $(q-1)^n \operatorname{Tr}(\psi \circ G) = \Sigma_{f(x) \not\equiv 0, x^{q-1}=1} G(x)$.

The main point in the proof of part (a) is that $F_{\varepsilon, \Delta, f}$ has an orthonormal basis consisting of elements of the form $k_{w,j} X^w / f^j$, where j runs over all of Z, X^w runs through a suitable set of monomials and the $k_{w,j}$ are constants of normalization. We make no attempt to reproduce this proof but will give some indication of how the trace formula (b) is deduced from (2). The idea is to prove (b) for the case in which G is a polynomial in X and $1/f$ and then take limits. With G so restricted choose r so large that $G f^{(q-1)p^r}$ is a polynomial and for $j \leqslant 0$ put

$$(\psi \circ G \circ f^{(q-1)p^r})(X^w / f^j) = \Sigma A_{(w,j),(u,k)} X^u / f^k .$$

The X^w/f^j with $j \leqslant 0$ form a basis of the space of polynomials and hence by (2),

$$\Sigma_{j \leqslant 0, w} A_{(w,j),(w,j)} = \Sigma_{x^{q-1}=1} \quad G(x) f(x)^{(q-1)p^r} / (q-1)^n \qquad (12)$$

which is close to $(q-1)^{-n} \Sigma_{\bar{f}(\bar{x}) \neq 0, x^{q-1}=1} G(x)$. Since $\psi \circ G$ is completely continuous, a good estimate for its trace can be obtained by considering the action of $\psi \circ G$ on elements of the form X^w/f^j such that $p^r > j$. But

$$(\psi \circ G)(X^w / f^j) = (\psi \circ G) f^{-p^r} X^w f^{p^r - j} = f^{-p^r} f^{p^r} (\psi \circ G) f^{-p^r} X^w f^{p^r - j}$$

and since $f(X^q)^{p^r}$ is in some sense close to f^{qp^r}, it follows that our last expression is close to

$$f^{-p^r} \psi (G f^{(q-1)p^r} X^w / f^{j-p^r}) = f^{-p^r} \Sigma A_{(w, j-p^r),(u,s)} X^u / f^s$$

and from this we conclude that $\operatorname{Tr}(\psi \circ G)$ is close to

$$\Sigma_{j \leqslant p^r} A_{(w, j-p^r),(w, j-p^r)} = \Sigma_{j \leqslant 0} A_{(w,j),(w,j)}$$

and the trace formula then follows from (12). This argument may be made precise.

By setting $G=1$, Theorem 2 gives an immediate proof of the meromorphic property of zeta functions. For L-series, REICH extended the result of SERRE by considering the case in which $\bar{E}/\bar{F}$ is cyclic of degree m prime to p and determining relations between the Artin symbol for $\bar{E}/\bar{F}$ and the Artin symbol for $\bar{E}'/\bar{F}'$, where $\bar{E}'=\bar{E}(\omega)$, $\bar{F}'=\bar{F}(\omega)$, ω is a primitive m^{th} root of unity. As in Chapter II, the Artin symbol for $\bar{E}'/\bar{F}'$ can be computed explicitly the improvement over the treatment of Theorem 1 being that we now have a trace formula which permits us to use a formula much simpler than equation (10). Put $G(X)=(f(X^q)/f(X))^{1/m}$, this lies in $F_{\varepsilon,\Delta,f}$ (for suitable ε, Δ) and for $\bar{f}(\bar{x})\neq 0$, $\bar{x}^q=\bar{x}$, the right side of equation (10) may be replaced by $qG(T(\bar{x}))$. By means of the relations between the two Artin symbols, REICH obtained the generalization of Theorem 1 stated in the introduction.

IV. L-series

We now sketch a somewhat different treatment of L-series based on the theory of dagger spaces of WASHNITZER-MUNSKY [7] and upon the connection found by KATZ [8] between this theory and our deformation theory [9, § 5].

For simplicity we shall for the most part restrict our attention to cyclic coverings of affine space but the arguments generalize to cyclic extensions of arbitrary function fields provided (as noted previously) one uses a special case of a general trace formula reported by MUNSKY. Except for this point we use only elementary parts of the theory of dagger spaces.

§ 1. If E is an abelian extension of a field F and E is the splitting field over F of $f\varepsilon F[t]$ having no multiple roots, then by the Euler Identities there exists a faithful representation, ϱ, of the galois group in $GL(d-1, F)$ given explicitly by the formula

$$\sigma\frac{1}{f'(\alpha)}\begin{pmatrix}1\\ \alpha\\ \vdots\\ \alpha^{d-2}\end{pmatrix}=\varrho(\sigma)\frac{1}{f'(\alpha)}\begin{pmatrix}1\\ \alpha\\ \vdots\\ \alpha^{d-2}\end{pmatrix}$$

for each $\sigma\varepsilon G(E/F)$, and each root α of f.

§ 2. As before let $\bar{K}=GF[q]$ and let K be a finite extension of the p-adic rationals whose residue class field is $\bar{K}$. If f is a polynomial in one variable, x, of degree d with coefficients in the ring of integers of K and if $\bar{f}$, the image of f in $\bar{K}[x]$, is of degree d with no multiple roots then we have associated [9] a space $\mathfrak{W}^S$ with f in the following way. Let $\bar{f}(X_1, X_2)=$

$X_2^d f(X_1/X_2)$. Let L be the space of power series in 3 variables X_0, X_1, X_2

$$L = \{\Sigma_{dw_0 = w_1 + w_2} B_w X^w \mid \operatorname{Inf}(\operatorname{ord} B_w - b w_0) > -\infty\},$$

where b is a suitably chosen positive real number and set

$$\mathfrak{W} = L/(D_1 L + D_2 L),$$

where

$$D_i = X_i \frac{\partial}{\partial X_i} + \pi X_0 X_i \frac{\partial \tilde{f}}{\partial X_i}, \quad (i = 1, 2)$$

and let $\mathfrak{W}^S$ be the image in $\mathfrak{W}$ of all elements of L which as power series are divisible by $X_1 X_2$.

It is known that $\{X_0 X_1^i X_2^{d-i}\}_{0<i<d}$ are elements of L whose image in $\mathfrak{W}^S$-span that space. Furthermore it is known that the operator

$$\alpha = \psi \circ \exp\{\pi X_0 \tilde{f}(X) - \pi X_0^q \tilde{f}(X^q)\}$$

induces on $\mathfrak{W}^S$ an endomorphism which determines the zeta function of the zero dimensional variety, $\bar{f}(x) = 0$.

Following Washnitzer and Monsky, we associate with f the space $(f)^\dagger$ consisting of all power series in x, $1/f'$,

$$\Sigma A_{ij} x^i f'^{-j} \qquad \operatorname{mod} f(x)$$

such that $\operatorname{Inf}(\operatorname{ord} A_{ij} - b(i+j)) > -\infty$ for some $b > 0$.
The Katz mapping Θ of $\mathfrak{W}^S$ into $(f)^\dagger$ is given explicitly by

$$X_0 X_1^i X_2^{d-i} \to x^{i-1}/f'(x), \qquad i = 1, 2 \ldots d-1 \tag{13}$$

More generally if $f_\Gamma(x)$ is a polynomial in one variable x parametrized rationally (over K) by $\Gamma = (\Gamma_1, \ldots, \Gamma_n)$, let $f'_\Gamma = \dfrac{\partial f_\Gamma}{\partial x}$, and let $R(\Gamma)$ be the resultant of f_Γ, f'_Γ (so the vanishing of $\bar{R}$ is the condition for multiplicity of roots) and let

$$\mathfrak{M} = \{\Gamma \in \Omega^n \mid |\Gamma_i| \leqslant 1, |R(\Gamma)| = 1\}$$

We suppose that $\mathfrak{M}$ is not empty. For each $\Gamma \in \mathfrak{M}$ we may as above form the space, $\mathfrak{W}_\Gamma^s$, and the important point is that the mapping

$$\alpha_\Gamma = \psi \circ \exp\{\pi X_0 \tilde{f}_\Gamma(X_1, X_2) - \pi X_0^q \tilde{f}_{\Gamma^q}(X_1^q, X_2^q)\}$$

induces a mapping, α_Γ^s, of $\mathfrak{W}_\Gamma^S$ into $\mathfrak{W}_{\Gamma^q}^S$ whose matrix, A_Γ, relative to the chosen bases, has coefficients which lie in a space of the type discussed in Chapter III, explicitly in $F_{\varepsilon, \Delta, R}$ for suitable $\varepsilon > 0$, $\Delta > 0$. On the other hand, for each $\Gamma \in \mathfrak{M}$ we may define the corresponding dagger ring, $f_\Gamma^\dagger$, and there exists a isomorphism τ_1 of $f_{\Gamma^q}^\dagger$ into $f_\Gamma^\dagger$ defined by

$$\tau_1 \Gamma = \Gamma$$
$$\tau_1 (x \bmod f_{\Gamma^q}) \equiv (x^q \bmod f_\Gamma) \bmod p \tag{14}$$

Letting τ be the inverse of τ_1 we know from KATZ [8] that the diagram

$$\begin{array}{ccc} f_\Gamma^\dagger & \overset{q\tau}{\to} & f_{\Gamma^q}^\dagger \\ \scriptstyle\theta_\Gamma \uparrow & & \uparrow \scriptstyle\theta_{\Gamma^q} \\ \mathfrak{W}_\Gamma^S & \underset{\alpha_\Gamma}{\to} & \mathfrak{W}_{\Gamma^q}^S \end{array}$$

commutes and hence putting

$$\mathfrak{X}_\Gamma = \frac{1}{f_\Gamma'(x)} \begin{pmatrix} 1 \\ x \\ \vdots \\ x^{d-2} \end{pmatrix} \bmod f_\Gamma(x),$$

a $d-1$ column vector with coefficients in $f_\Gamma^\dagger$, we have.

$$q\tau \mathfrak{X}_\Gamma = A_\Gamma \mathfrak{X}_{\Gamma^q} \tag{15}$$

If $\Gamma_0^q = \Gamma_0$, $\Gamma_0 \in \mathfrak{M}$, then the splitting field, E_{Γ_0}, of f_{Γ_0} over K is unramified over K and the mapping τ_1 specializes to the Frobenius of E_{Γ_0} over K. More generally, if $\Gamma_0^{q^s} = \Gamma_0$, $\Gamma_0 \in \mathfrak{M}$ then put $F_{\Gamma_0} = K(\Gamma_0)$, let E_{Γ_0} be the splitting field of f_{Γ_0} over F_{Γ_0} then for each root α of f_{Γ_0} we may conclude that if s is a multiple of $\deg F_{\Gamma_0}/K$ then

$$(\text{Frobenius of } E_{\Gamma_0}/F_{\Gamma_0})^{-s/\deg(F_{\Gamma_0}/K)} \frac{1}{f_{\Gamma_0}'(\alpha)} \begin{pmatrix} 1 \\ \alpha \\ \vdots \\ \alpha^{d-2} \end{pmatrix} \tag{16}$$
$$= \frac{1}{q^s} A_{\Gamma_0} A_{\Gamma_0^q} \ldots A_{\Gamma_0^{q^{s-1}}} \frac{1}{f_{\Gamma_0}'(\alpha)} \begin{pmatrix} 1 \\ \alpha \\ \vdots \\ \alpha^{d-2} \end{pmatrix}$$

Finally the matrix A_Γ may be connected with classical things. For simplicity of exposition we assume that $|R(0)| = 1$ and then if for Γ close to 0 we think of $\mathfrak{X}_\Gamma$ as referring to a locally holomorphic function of Γ (i.e. $x \bmod f_\Gamma(x)$ is a root of the polynomial viewed as a function of Γ for Γ close to 0), then we may define a matrix, C_Γ, holomorphic near 0 such that

$$\mathfrak{X}_\Gamma = C_\Gamma \mathfrak{X}_0 . \tag{17}$$

C_Γ has coefficients which are formal power series in Γ and which do not depend on whether $\mathfrak{X}_\Gamma$ is viewed from the classical or p-adic point of view. However for Γ close to 0 in the p-adic sense, we have

$$A_\Gamma = C_\Gamma A_0 C_{\Gamma^q}^{-1} \tag{18}$$

and this specification of A_Γ for Γ near 0 determines A_Γ for all $\Gamma \in \mathfrak{M}$. Finally we note that $\mathfrak{X}_\Gamma$ satisfies a set of linear first order partial differential equations with rational coefficients and hence C_Γ satisfies the same equations.

These differential equations together with the initial condition $C_0 = I$ in fact determine C_Γ for Γ close to 0.

Note: It is no doupt possible to obtain the main results (16), (17), (18) and the holomorphic nature of A_Γ directly from the theory of Washnitzer-Monsky and without the application of our deformation theory.

§ 3. Now let $\bar{U}$ be an abelian covering of degree d of a variety, $\bar{V}$, defined over $\bar{K}$. For simplicity let $\bar{V}$ be affine n space with generic point $\bar{\Gamma} = (\bar{\Gamma}^{(1)}, \ldots, \bar{\Gamma}^{(n)})$. Let $\bar{F}_\Gamma = \bar{K}(\bar{\Gamma})$, the field of functions of $\bar{V}$ and let $\bar{E}_\Gamma$ be the function field of $\bar{U}$. We may lift $\bar{F}_\Gamma$ (resp: $\bar{E}_\Gamma$) to F_Γ (resp: E_Γ), an algebraic extension of $K(\Gamma)$, the field of rational functions in n indeterminates $\Gamma = (\Gamma^{(1)}, \ldots, \Gamma^{(n)})$, so that E_Γ is a galois extension of F_Γ with an isomorphism, φ, of $\mathfrak{G}(\bar{E}_\Gamma/\bar{F}_\Gamma)$ onto $\mathfrak{G}(E_\Gamma/F_\Gamma)$ defined in the usual way. Thus if ν is the place of E_Γ defined by reduction mod p then

$$\nu \circ \sigma = \varphi^{-1}(\sigma) \circ \nu \tag{19}$$

for each $\sigma \in \mathfrak{G}(E_\Gamma/F_\Gamma)$. Let $\bar{\beta}$ be a generator of $\bar{E}_\Gamma$ over $\bar{F}_\Gamma$, let β be an element of E_Γ such that $\nu(\beta) = \bar{\beta}$ and let $\bar{f}_\Gamma$ (resp: f_Γ) be the irreducible polynomial over $\bar{F}_\Gamma$ (resp: F_Γ) satisfied by $\bar{\beta}$ (resp: β). In the present section and again in § 4 we impose the restrictive hypothesis that in the above construction we may take $F_\Gamma = K(\Gamma)$. This restriction is removed in § 5.

If Γ_0 is an element of Ω^n which is algebraic over K, then the specialization $\Gamma \to \Gamma_0$ gives a place of F_Γ which can be extended (non-uniquely) to a place, $\mathfrak{p}$, of E_Γ taking values in Ω. Let E_{Γ_0} be the residue class field of E_Γ (i.e. the image of E_Γ under $\mathfrak{p}$). The field, E_{Γ_0}, is independent of the choice of $\mathfrak{p}$ and is the splitting field of f_{Γ_0} over $F_{\Gamma_0} (= K(\Gamma_0))$. If $R(\Gamma_0) \neq 0$ then the valuation $\mathfrak{p}$ is unramified over F_Γ and hence $\mathfrak{G}(E_{\Gamma_0}/F_{\Gamma_0})$ is naturally isomorphic to $\mathfrak{G}_\mathfrak{p}$ the $\mathfrak{p}$ decomposition subgroup of $\mathfrak{G}(E_\Gamma/F_\Gamma)$. Since $\mathfrak{G}(E_\Gamma/F_\Gamma)$ is abelian, both $\mathfrak{G}_\mathfrak{p}$ and the injection of $\mathfrak{G}(E_{\Gamma_0}/F_{\Gamma_0})$ into $\mathfrak{G}(E_\Gamma/F_\Gamma)$ is independent of the choice of $\mathfrak{p}$ (and hence may be denoted by the symbol η_{Γ_0}).

Similarly if $\bar{\Gamma}_0$ is algebraic over $\bar{K}$, then corresponding to a place, $\bar{\mathfrak{p}}$, extending the specialization $\bar{\Gamma}\to\bar{\Gamma}_0$ we have an injection η_{Γ_0} of $\mathfrak{G}(\bar{E}_{\Gamma_0}/\bar{F}_{\Gamma_0})$ into $\mathfrak{G}(\bar{E}_\Gamma/\bar{F}_\Gamma)$ provided $\bar{R}(\bar{\Gamma}_0)\neq 0$.

Finally if $\Gamma_0\in\mathfrak{M}$ (cf. § 2) then $E_{\Gamma_0}/F_{\Gamma_0}$ is unramified and hence its galois group is naturally isomorphic to that of $\bar{E}_{\Gamma_0}/\bar{F}_{\Gamma_0}$. We denote this isomorphism by φ_{Γ_0}.

By standard arguments we obtain a relation between these four types of maps.

Theorem 3. If $\Gamma_0\in\mathfrak{M}$ is algebraic over K then the diagram

$$\begin{array}{ccc}
\mathfrak{G}(E_\Gamma/F_\Gamma) & \underset{\varphi}{\leftarrow} & \mathfrak{G}(\bar{E}_\Gamma/\bar{F}_\Gamma)\\
\eta_{\Gamma_0}\big\uparrow & & \big\uparrow\eta_{\Gamma_0}\\
\mathfrak{G}(E_{\Gamma_0}/F_{\Gamma_0}) & \underset{\varphi_{\Gamma_0}}{\leftarrow} & \mathfrak{G}(\bar{E}_{\Gamma_0}/\bar{F}_{\Gamma_0})
\end{array}$$

commutes.

If $R(\Gamma_0)\neq 0$ and if σ_0 (resp: σ) is an element of $\mathfrak{G}(E_{\Gamma_0}/F_{\Gamma_0})$ (resp: $\mathfrak{G}(E_\Gamma/F_\Gamma)$) then by the methods of § 1, using f_{Γ_0} (resp: f_Γ), define B_{σ_0} (resp: $B^{(\sigma)}(\Gamma)$) in $GL(d-1, F_{\Gamma_0})$ (resp: $GL(d-1, F_\Gamma)$) such that for each root α of f_{Γ_0} (resp: f_Γ)

$$\sigma_0\frac{1}{f'_{\Gamma_0}(\alpha)}\begin{pmatrix}\overset{1}{\alpha}\\ \vdots\\ \alpha^{d-2}\end{pmatrix}=B_{\sigma_0}\frac{1}{f'_{\Gamma_0}(\alpha)}\begin{pmatrix}\overset{1}{\alpha}\\ \vdots\\ \alpha^{d-2}\end{pmatrix}\tag{20}$$

$$\left(\text{resp:}\quad \sigma\frac{1}{f'_\Gamma(\alpha)}\begin{pmatrix}\overset{1}{\alpha}\\ \vdots\\ \alpha^{d-2}\end{pmatrix}=B^{(\sigma)}(\Gamma)\frac{1}{f'_\Gamma(\alpha)}\begin{pmatrix}\overset{1}{\alpha}\\ \vdots\\ \alpha^{d-2}\end{pmatrix}\right)$$

Theorem 4. If $R(\Gamma_0)\neq 0$ and $\sigma_0\in\mathfrak{G}(E_{\Gamma_0}/F_{\Gamma_0})$ then $\sigma=\eta_{\Gamma_0}(\sigma_0)$ if and only if

$$B_{\sigma_0}=B^{(\sigma)}(\Gamma_0).$$

We apply this result to obtain information concerning a image of the Artin symbol of $\bar{E}_\Gamma/\bar{F}_\Gamma$ under a representation of the galois group. We define a faithful representation, ϱ, of $\mathfrak{G}(\bar{E}_\Gamma/\bar{F}_\Gamma)$ in $GL(d-1, F_\Gamma)$ by setting

$$\varrho(\bar{\sigma})=B^{\varphi(\bar{\sigma})}$$

for each $\bar{\sigma}\in\mathfrak{G}(\bar{E}_\Gamma/\bar{F}_\Gamma)$.

If now $\Gamma_0=\Gamma_0^q$, $\Gamma_0\in\mathfrak{M}$, let $\bar{\sigma}_0$ be the Frobenius of $\bar{E}_{\Gamma_0}/\bar{F}_{\Gamma_0}$ then $\eta_{\Gamma_0}(\bar{\sigma}_0)=(\bar{\Gamma}_0, \bar{E}_\Gamma/\bar{F}_\Gamma)$ and letting σ denote $\varphi((\bar{\Gamma}_0, \bar{E}_\Gamma/\bar{F}_\Gamma))$, we have

$$\varrho((\Gamma_0, \bar{E}_\Gamma/\bar{F}_\Gamma))=B^{(\sigma)}(\Gamma)$$

On the other hand since $\sigma=(\varphi\circ\eta_{\Gamma_0})(\bar{\sigma}_0)$, we see by Theorem 3 that $\sigma=\eta_{\Gamma_0}(\varphi_{\Gamma_0}(\bar{\sigma}_0))$, while clearly $\varphi_{\Gamma_0}(\bar{\sigma}_0)$ is the Frobenius of $E_{\Gamma_0}/F_{\Gamma_0}$ (which for the moment we denote by σ_0). Hence by § 2, $(B_{\sigma_0})^{-1}=q^{-1}A_{\Gamma_0}$. Finally, since $\sigma=\eta_{\Gamma_0}(\sigma_0)$, Theorem 4 shows that $B^{(\sigma)}(\Gamma_0)=(q^{-1}A_{\Gamma_0})^{-1}$. Thus under the stated hypothesis,

$$\varrho((\bar{\Gamma}_0, \bar{E}_\Gamma/F_\Gamma))^{-1} = B^{(\sigma)}(\Gamma)$$

where

$$B^{(\sigma)}(\Gamma_0) = q^{-1}A_{\Gamma_0}$$

More generally:

Theorem 5. If

$$\Gamma_0^{q^s} = \Gamma_0, \Gamma_0 \in \mathfrak{M}$$

then

$$\varrho((\bar{\Gamma}_0, \bar{E}_\Gamma/\bar{F}_\Gamma)^{-s/\deg(\bar{F}_{\Gamma_0}/\bar{K})}) = B^{(\sigma)}(\Gamma)$$

where

$$B^{(\sigma)}(\Gamma_0) = q^{-1}A_{\Gamma_0}q^{-1}A_{\Gamma_0^q}\dots q^{-1}A_{\Gamma_0^{q^{s-1}}}$$

§ **4.** We now suppose E_Γ/F_Γ cyclic of degree d. Let ω be a primitive d^{th} root of unity in Ω. We simultaneously diagonalize $B^{(\sigma)}(\Gamma)$ for all $\sigma\in\mathfrak{G}(E_\Gamma/F_\Gamma)$. To do this let $E'_\Gamma=E_\Gamma(\omega)$, $F'_\Gamma=F_\Gamma(\omega)$ and with the notation of § 3, $E_\Gamma=F_\Gamma(\beta)$, $E'_\Gamma=F'_\Gamma(\beta)$. Let γ be any other generator of E'_Γ over F'_Γ and let $g\in F'_\Gamma[X]$ be the irreducible polynomial over F'_Γ satisfied by γ.

Let

$$S=\left\{\frac{1}{f'_\Gamma(\beta)}, \frac{\beta}{f'_\Gamma(\beta)}, \dots, \frac{\beta^{d-2}}{f'_\Gamma(\beta)}\right\}, \quad S'=\left\{\frac{1}{g'(\gamma)}, \frac{\gamma}{g'(\gamma)}, \dots, \frac{\gamma^{d-2}}{g'(\gamma)}\right\}.$$

The Euler identities show that S and S' span the same space, V, over F_Γ.

If each $\sigma\in\mathfrak{G}(E_\Gamma/F_\Gamma)$ is extended to E'_Γ by requiring σ to be the identity map of F'_Γ then the matrix $B^{(\sigma)}(\Gamma)$ represents (relative to the basis S) the action of σ as F'_Γ linear endomorphism of V. If we let $D^{(\sigma)}(\Gamma)$ be the matrix of σ as F'_Γ-linear endomorphism of V relative to the basis S' then there exists $H\in GL(d-1, F'_\Gamma)$, independent of σ such that

$$D^{(\sigma)}(\Gamma) = H(\Gamma)B^{(\sigma)}H(\Gamma)^{-1} \tag{21}$$

Since E'_Γ is a Kummer extension of F'_Γ, we may choose g to be

$$g(X) = X^d - a \tag{22}$$

for suitable $a\in F'_\Gamma$. Since $\gamma^i/g'(\gamma)=\gamma^i/d\gamma^{d-1}$, it is clear that $D^{(\sigma)}(\Gamma)$ is not only diagonal, but also has coefficients independent of Γ. Thus we obtain a representation, ϱ', of $\mathfrak{G}(\bar{E}_\Gamma/\bar{F}_\Gamma)$ in $GL(d-1, Q(\omega))$ defined by

$$\varrho'(\bar{\sigma}) = D^{(\varphi(\bar{\sigma}))} = H\varrho(\bar{\sigma})H^{-1} \tag{23}$$

and this representation is clearly a direct sum of the $d-1$ non-trivial linear characters of the galois group.

Under the hypothesis of Theorem 5 we now have,

$$\varrho'((\bar{\Gamma}_0, \bar{E}_\Gamma/\bar{F}_\Gamma)^{-s/\deg(\bar{F}_{\Gamma_0}/\bar{K})}) = D^{(\sigma)}(\Gamma) = H(\Gamma)B^{(\sigma)}(\Gamma)H(\Gamma)^{-1},$$

where $B^{(\sigma)}(\Gamma_0)$ is given by Theorem 5. If now $H(\Gamma_0)$, $H(\Gamma_0)^{-1}$ are well defined, then since $D^{(\sigma)}(\Gamma)$ is independent of Γ, we have

$$D^{(\sigma)}(\Gamma) = D^{(\sigma)}(\Gamma_0) = H(\Gamma_0)\frac{1}{q}A_{\Gamma_0}\cdots\frac{1}{q}A_{\Gamma_0^{q^{s-1}}}H(\Gamma_0)^{-1}$$

and hence if $H(\Gamma_0)$, $H(\Gamma_0^q), \ldots, H(\Gamma_0^{q^{s-1}})$ and their inverses are all well defined then

$$D^{(\sigma)}(\Gamma) = B_{\Gamma_0}B_{\Gamma_0^q}\cdots B_{\Gamma_0^{q^{s-1}}}$$

where

$$B_\Gamma = q^{-1}H(\Gamma)A_\Gamma H(\Gamma^q)^{-1}.$$

We now reformulate Theorem 5. We note that the polar locus of H and H^{-1} need not be defined over K but by taking a finite number of the conjugates over K of these loci we obtain a locus defined over K which we shall refer to as the K polar locus of H and H^{-1}.

Theorem 6. If $\Gamma_0^{q^s} = \Gamma_0$, $\Gamma_0 \in M$, Γ_0 not on the K polar locus of either H or H^{-1} then

$$\varrho'((\bar{\Gamma}_0, \bar{E}_\Gamma/\bar{F}_\Gamma)^{-s/\deg(\bar{F}_{\Gamma_0}/\bar{K})}) = B_{\Gamma_0}B_{\Gamma_0^q}\cdots B_{\Gamma_0^{q^{s-1}}} \tag{24}$$

where B_Γ is a matrix with coefficients in a space of functions of the type studied by Reich.

If now B_Γ is diagonal then (24) would remain valid with ϱ' replaced by an arbitrary linear character of $\mathfrak{G}(\bar{E}_\Gamma/\bar{F}_\Gamma)$ and B_Γ replaced by a function of the type to which the Reich trace formula may be applied. Comparing equation (24) with equation (7) we see that we would then have the correct formula to prove meromorphy of L-series. Thus to complete the proof (at least for F_Γ a purely transcendental field) it is enough to show that B_Γ is diagonal. To simplify the argument we assume that 0 is not on the polar locus of $H(\Gamma)$ or $H(\Gamma)^{-1}$.

Theorem 7. The matrix B_Γ is diagonal

Proof. Since $D^{(\sigma)}(\Gamma)$ is diagonal it is clear that B_0 is diagonal and using § 2, we see that for Γ close to 0,

$$B_\Gamma = T_\Gamma B_0 T_{\Gamma^q}^{-1}$$

where

$$T_\Gamma = H(\Gamma)\, C_\Gamma H(0)^{-1} \tag{25}$$

Since $T_0 = I$, it is enough to show that $T_\Gamma^{-1} \dfrac{\partial T_\Gamma}{\partial \Gamma^{(i)}}$ is diagonal for $i = 1, 2, \ldots, n$.

Now equation (17) and (25) show that the differential equations satisfied by T_Γ are precisely the same as those satisfied by

$$\mathfrak{X}'_\Gamma = \frac{1}{g'(x)} \begin{pmatrix} 1 \\ x \\ \vdots \\ x^{d-2} \end{pmatrix} \bmod g(x) = H(\Gamma)\, \mathfrak{X}_\Gamma \tag{26}$$

and a routine computation, using (22) shows that

$$\frac{\partial}{\partial \Gamma^{(i)}} \mathfrak{X}'_\Gamma = \frac{1}{a} \frac{\partial a}{\partial \Gamma^{(i)}} \cdot d^{-1} \begin{pmatrix} 1-d & 0 & \ldots\ldots & 0 \\ 0 & 2-d & \ldots\ldots & 0 \\ \cdots & & \ldots\ldots\ldots\ldots & \cdot \\ 0 & 0 & \ldots\ldots\ldots & -1 \end{pmatrix} \mathfrak{X}'_\Gamma . \tag{27}$$

This completes the proof of the theorem.

§ 5. We now briefly indicate how the preceeding paragraphs should be modified if $\bar{F}_\Gamma$ is not a purely transcendental extension of $\bar{K}$. We may suppose $\bar{F}_\Gamma$ is the field of functions of a hypersurface,

$$\bar{h}(\bar{\Gamma}^{(1)}, \ldots, \bar{\Gamma}^{(n)}) = 0, \tag{28}$$

defined over $\bar{K}$ and we may further suppose that $\partial\bar{h}/\partial\bar{\Gamma}^{(n)}$ is not everywhere zero on this hypersurface. We lift $\bar{h}$ to h, a polynomial with coefficients in K and lift $\bar{F}_\Gamma$ to F_Γ, the field of functions of the lifted hypersurface. We define $h^\dagger$ to be the space of power series in $\Gamma^{(1)}, \Gamma^{(2)}, \ldots, \Gamma^{(n)}, (\partial h/\partial\Gamma^{(n)})^{-1}$ mod h with growth conditions customary for dagger spaces, and as in § 2, construct a lifting, μ, of the Frobenius to $h^\dagger$, i.e. a ring endomorphism of $h^\dagger$ such that

$$\mu\Gamma^{(i)} \equiv (\Gamma^{(i)})^q \bmod p, \qquad i = 1, 2, \ldots, n. \tag{29}$$

(To be explicit, we may set $\mu\Gamma^{(i)} = (\Gamma^{(i)})^q$ for $i = 1, 2, \ldots, n-1$, and we may extend $\mu\Gamma^{(n)}$ to a function of type $F_{\varepsilon,\Delta,\partial h/\partial\Gamma}(n)$ (cf. chapter III) such that $h(\mu\Gamma) = h(\Gamma)^q$). It is easily checked that if for Γ close to 0 we set (cf. § 2 above)

$$A'_\Gamma = C_\Gamma A_0 C_{\mu\Gamma}^{-1} \tag{30}$$

(supposing for simplicity that $\partial h/\partial\Gamma^{(n)}$ does not vanish at $\Gamma=0$) then A'_Γ is a matrix whose coefficients are functions of the type $F_{\varepsilon,\Delta,R_1}$, where $R_1 = R\partial h/\partial\Gamma^{(n)}$. If we replace A_Γ in § 2 by A'_Γ then the remainder of the argument causes no difficulty and the procedure of § 4 now reduces the rationality of L-series to the evaluation of sums of the form

$$M_s = \Sigma G(\Gamma)\, G(\mu\Gamma) \ldots G(\mu^{s-1}\Gamma) \tag{31}$$

the sum being over $\{\Gamma | \mu^s\Gamma = \Gamma,\, h(\Gamma)=0,\, R_1(\Gamma) \not\equiv 0 \bmod p\}$ and where $G \in F_{\varepsilon,\Delta,R_1}$ for suitable ε, $\Delta > 0$.

Associated with the pair (h, R_1) there is a dagger space of power series in Γ, R_1^{-1} mod h with suitable growth conditions. Denoting this space by the symbol (h, R_1), and following Munsky, the mapping ψ, may be generalized by defining the relative trace of (h, R_1) over $\mu(h, R_1)$ and composing this on the left with μ^{-1}, where the symbol μ is now used to denote the lifting of the Frobenius to (h, R_1). Denoting this operation by ψ, we know from Monsky that $\psi \circ G$ is a completely continuous endomorphism of (h, R_1), with trace formula adequate for the task of showing exp $\{\Sigma M_s t^s/s\}$ is p-adically meromorphic.

Bibliography

[1] Dwork, B.: Amer. J. Math. **82**, 631–648 (1960).
[2] Serre, J. P.: Séminaire Bourbaki **198** (1959–1960).
[3] Reich, D.: Princeton Thesis 1966.
[4] Serre, J. P.: Pub. Math. IHES Paris **12** (1962).
[5] Warning, E.: Abh. Math. Sem. Hamburg **11**, 76–83 (1935).
[6] Dwork, B.: Journal für die Reine und angewandte Math. **23**, 130–142 (1960).
[7] Monsky, P. and G. Washnitzer, Proc. Nat. Acad. Sci (USA) **52**, 1511–1514 (1964).
[8] Katz, N.: Princeton Thesis 1966.
[9] Dwork, B.: Annals of Math., **80**, 227–299 (1964).
[10] Grothendieck, A.: Seminaire Bourbaki **279** (1964–1965).

Linear Topological Spaces over Non-Archimedean Valued Fields

A. F. Monna

Introduction

I propose to give a survey of the results of the study of linear topological spaces over non-archimedean valued fields. Proofs of theorems will not be given.

The first results on this subject go back to 1943 [*1a*], [*1b*]. At first a theory of non-archimedean normed spaces was attempted [*2*]. In more recent years a theory of locally convex spaces over non-archimedean valued fields followed.

Both parts of the theory are now in development. Several problems, which have found a solution in spaces over the reals, still await a solution in our case. Nevertheless, as a general conclusion it may be said, that many parts of the classical theory remain valid. It is remarkable that this is also true of parts for which one would expect the ordering of the reals to be essential. I mention, for instance, the separation theorems for convex sets; without using an ordering of the fields – even if ordering should be possible – one can define convexity of sets and prove separation theorems for convex sets. In many cases the proofs which are valid for the real spaces, cannot be given in the same way for spaces over a non-archimedean valued field K. Thus, a way of unifying both theories is desirable.

By K I shall always denote a non-archimedean valued field. For reasons of simplicity I suppose that K is complete and that the valuation is not trivial. Neither of the two conditions is necessary for all results, but for a general review of the field it suffices to consider this case.

In the first part I consider normed spaces; in the second I shall consider the problems of convexity. In the third part I give a brief review of applications of the general theory. The literature at the end has not the pretention of being complete. I also mention some literature concerning these problems in spaces over the reals which seem to be relevant for comparison with the theory of spaces over a field K.

§ 1. Non-archimedean normed spaces

1.1. A normed linear space E over K is a non-archimedean normed space if the norm $\|.\|$ satisfies

$$\|x+y\| \leqslant \max(\|x\|, \|y\|), \qquad x, y \in E.$$

It follows that

$$\|x+y\| = \max(\|x\|, \|y\|) \quad \text{if} \quad \|x\| \neq \|y\|.$$

E is a non-archimedean Banach space if E is complete.

Examples

(i) The space of all bounded sequences $x=(x_i)_{i\in N}$, $x_i \in K$, normed by $\|x\| = \sup_i |x_i|$.

(ii) Let X be a locally compact topological space and $C_c(X)$ the linear space of the continuous functions with compact support $f: X \to K$. It is reasonable to require that the functions of $C_c(X)$ separate the points of X; it can be proved that this is the case if and only if X is zero-dimensional. If we put $\|f\| = \sup_{x \in X} |f(x)|$, $C_c(X)$ is a non-archimedean Banach space.

(iii) In a more general way one can consider normed linear spaces over a valuated field and ask for necessary and sufficient conditions for the norm $\|.\|$ being non-archimedean. It is evidently necessary that the valuation of K is non-archimedean, but this condition is not sufficient. The linear space l^1 of all sequences $x=(x_1, x_2, \ldots)$, $x_i \in K$, such that $\sum |x_i| < \infty$, topologized by means of the norm

$$\|x\| = \sum_i |x_i|$$

is a counterexample. This norm does not satisfy the strong inequality (see [*1b*]).

Linear functions are defined, as usual, as continuous linear mappings from $E \to K$. The relation between continuity and boundedness remains valid. However, the Hahn-Banach theorem is valid only under certain restrictions on K. Therefore I have to introduce the notion of spherical completeness; this notion is important throughout the whole theory.

I enlarge the conditions somewhat because this is useful for our later considerations on locally convex spaces. I shall suppose that the topology on E is determined by a semi-norm p, that is to say, I require $p(x) \geqslant 0$, $p(0)=0$, $p(ax)=|a|\,p(x)$ for all $a \in K$ and $p(x+y) \leqslant \max(p(x), p(y))$. I denote a linear space over K, topologized by means of a seminorm p, if this is useful by (E, p). The set $\{x \mid p(x-x_0) \leqslant \lambda\}$ will be called a sphere of centre x_0 and radius λ. Spheres have remarkable properties, for instance every point of the sphere may be taken as centre and if the intersection of two spheres is not empty, one of the spheres contains the other. A collection $\mathscr{C}$ of sets will be said to have the binary intersection property if every subcollection of $\mathscr{C}$, any two members of which intersect, has a non-void intersection. The space (E, p) will be called *spherically complete* [*3*], if the collection of its spheres has the binary intersection property. It is the same thing to say that any set of spheres which is totally ordered by inclusion, has a non-empty intersection. Spherical completeness implies completeness, but the converse is not true. I believe that the real meaning of this notion with regard to the problems in the spaces (E, p) is not yet quite clear. What happens, for instance, if in this definition the spheres are replaced by convex sets (see the definition of convexity in § 2)? I also mention the problem of the embedding of a space in a spherically complete space (see CARPENTIER [*4*]).

The definition of spherical completeness applies to the case of a valued field K, considered as a space over itself. In that case the following properties are equivalent:

(i) K is spherically complete.

(ii) Every pseudo-convergent sequence [*5*] in K has a pseudo-limit.

(iii) K is maximally complete in the sense that K cannot be extended without extending the value group or the residue class field. See for these equivalences KAPLANSKY [*6*], LAZARD [*7*] and VAN TIEL [*14*]. Complete fields with discrete valuation are spherically complete.

1.2. I now come to the theorem of Hahn-Banach. In the real or complex case the theorem concerns the extension of linear continuous functions; it is well known that the general problem of the extension of linear maps is in this case much more complicated.

The space (E, p) is said to have the *extension property* if it possesses the following property: for every space (F, q) over K and any linear subspace $V \subset F$, any linear map $\varphi: V \to E$ for which $p(\varphi(x)) \leqslant q(x)$, $x \in V$, has a linear extension $\psi: F \to E$ satisfying $p(\psi(x)) \leqslant q(x)$, $x \in F$.

The problem is to find necessary and sufficient conditions for a space having the extension property. For normed spaces over the reals the validity of the extension property for maps is a severe condition. NACHBIN [8] has given a solution of this problem for spaces over the reals and he gives a characterization of the spaces having the extension property in terms of function spaces.

For spaces over fields K with non-archimedean valuation some authors have given sufficient conditions (MONNA, COHEN [9]). INGLETON [3] has given the following necessary and sufficient condition: *(E, p) has the extension property only and only if (E, p) is spherically complete.*

Taking for E the field K as a space over itself, one gets the classical Hahn-Banach theorem for the non-archimedean case. But in our case the condition applies to more general situations. It is likely that this fact may be useful in functional analysis over non-archimedean valued fields, in particular for local fields, which indeed are spherically complete.

I need not enter into the immediate consequences from the Hahn-Banach theorem, for instance the existence of linear continuous functions. I give some examples.

(i) Let E be complete. Put $N_p = \{p(x) | x \in E\}$. Then (E, p) is spherically complete if in N_p every decreasing non-stationary sequence converges to 0.

Evidently, this condition can only be satisfied if the valuation of K is discrete. However, the condition does not imply that N_p is discrete in R_+^*.

In particular, the extension property is true for (E, p) if 0 is the only point of accumulation of N_p.

In a field K this condition is equivalent to the condition that the valuation of K is discrete. It should be noted however, that there exist fields with dense valuation which are maximally complete; for these fields the extension property is valid.

(ii) The Hahn-Banach theorem is valid for any field K with discrete valuation.

It is here the place to mention an important difference with the theory over the reals. Putting $N_K = \{|a| \; | a \in K\}$, the relation $\|ax\| = |a|.\|x\|$ implies $N_K N_p \subset N_p$. However, the relation $N_K N_p = N_K$, which is true if $K = R$, is not true in general. Let $S \subset R_+$ and $N_K S \subset S$; it can then be proved that there exist spaces (E, p) for which $N_p = S$. A consequence of this situation is, for instance, that it is not always possible to transform a given vector into a vector of norm 1 by means of a homothetic transformation.

1.3. An important subject in the study of the structure of normed spaces is the problem of the existence of a basis. The conditions under which such a basis exist give in our case better results than for real spaces.

Let E be a non-archimedean normed space over K. The problem is to give necessary and sufficient conditions under which there is in E a set $(e_i)_{i\in I}$ such that each $x\in E$ can be written in a unique way as a series

$$x = \sum_{i=1}^{\infty} x_i e_i , \qquad x_i \in K ,$$

where the convergence of the series is meant in the sense of convergence in the norm, and such that

$$\|x\| = \sup \|x_i e_i\| .$$

If such a set exists, it is called an *orthogonal basis.* A certain geometrical interpretation can be given of the relation which is required for $\|x\|$. A set $X\subset E-\{0\}$ is said to have the property (π) if for each set $(\lambda_x)_{x\in X}$, $\lambda_x\in K$, such that $\lambda_x=0$ but for a finite number

$$\|\sum_{x\in X} \lambda_x x\| = \sup_{x\in X} \|\lambda_x x\| .$$

It can then be proved that X has the property (π) if and only if for each collection $e_1,\ldots, e_n\in X$, $\lambda_1,\ldots, \lambda_n\in K$ and $e\in X-\{e_1,\ldots, e_n\}$

$$\|e - \sum \lambda_i e_i\| \geqslant \|e\| .$$

This means that $\|e\|$ is the shortest distance of e to the linear subspace determined by the set (e_i) and thus (π) means a kind of orthogonality. An orthogonal basis is a total system which has the property (π).

The existence of such a basis has been proved in the following cases [*10*].

1. *Let the valuation of K be discrete. Suppose that each strictly decreasing subset of the set $\{\|x\| \mid x\in E\}$ converges to* 0. *Then E has an orthogonal basis.*

2. *Let K be spherically complete. Suppose there is in E a dense subspace with finite or denumerable dimension. Then E has an orthogonal basis.*

CARPENTIER [*4*] proved a general theorem from which these theorems can be derived. In a certain sense these results are the best possible. I mention the following theorem: *Let E be an infinite dimensional Banach space and suppose that the norm does not satisfy the condition expressed in 1. Then, if E has an orthogonal basis, E is not spherically complete.*

Now, let l^∞ be the non-archimedean Banach space over K of the bounded sequences $(\lambda_i)_{i\in I}$, $\lambda_i\in K$, where the norm is given by

$$\|(\lambda_i)_{i\in I}\| = \sup_{i\in I} |\lambda_i| .$$

Suppose that K is spherically complete. It is then easily proved that this space is spherically complete. Applying the theorem to these spaces, one gets:

Suppose that the valuation of K is not discrete; let K be spherically complete. Then no infinite dimensional space l^∞ over K has an orthogonal basis.

The existence theorems of orthogonal bases are useful in the study of spaces over K. I mention, for instance, the problem of reflexivity; I refer to this later on in connection with convexity.

§ 2. Convexity

2.1. Let E be a linear space over K. A set $C \subset E$ is called K-convex if $x \in C$, $y \in C$ imply $\lambda x + \mu y \in C$ for all $\lambda, \mu \in K$, $|\lambda| \leqslant 1$, $|\mu| \leqslant 1$ (that is if C is a module over the ring of the integers of K) or if C can be transformed by a translation into a set having this property [11a]. An equivalent definition is (SPRINGER [*12*]): C is K-convex if $x, y, z \in C$ imply $\lambda x + \mu y + \nu z \in C$ for all integers λ, μ, ν in K such that $\lambda + \mu + \nu = 1$.

It is easily seen that K-convex sets have many of the usual properties of convex sets over the reals. Absorbing sets are defined in the customary way [*13*]. As in real spaces a semi-norm p corresponds to a convex absorbing set. This semi-norm is non-archimedean, but is it not uniquely determined: there is a minimal and a maximal semi-norm. The reason for this lies in the following fact. If E is topologized by means of p, the sets $\{x \in E \mid p(x) \leqslant 1\}$ and $\{x \in E \mid p(x) < 1\}$ are both open and closed; therefore the boundary of these sets is empty. E is zero-dimensional. This fact gives rise to some difficulties in the theory.

Locally K-convex spaces over K can now be defined in the usual way [*11b*] and a theory of these spaces can be developed. In particular, if K is spherically complete, the existence of linear continuous functions $\neq 0$ can be proved. The space l^1, introduced in § 1, example (iii), is a linear topological space over K which is not locally K-convex when convexity is defined as is done here. It is an open question whether there exist linear topological spaces over a field K with non-archimedean valuation in which all linear continuous functions are 0.

A duality theory for locally K-convex spaces was given by VAN TIEL [*14*]. As a remarkable fact I mention that there exist reflexive spaces, but these spaces cannot be normed unless they are finite dimensional. Indeed, the following theorem can be proved (see the article by MONNA and SPRINGER, quoted in [*10*]): *suppose that K is spherically complete; let E be a reflexive Banach space over K; then the dimension of E is finite.* Earlier on it was already proved that infinite dimensional normed spaces which have an orthogonal basis are not reflexive. Comparing this situation with the theory of reflexivity in real spaces, it must be remarked that several problems still await solution in our case. I mention, for instance, the following criterion for reflexivity, due to KLEE for spaces over the reals: a normed linear space over R is reflexive if and only if every decreasing sequence of closed bounded convex sets has a non-empty intersection [*15*].

SPRINGER [*12*] introduced the notion of c-compact set, which proved to be useful for duality theory. Let E be a vector space over K; let $X \subset E$. A filter $\mathcal{F}$ on X is called a convex filter if $\mathcal{F}$ has a basis which consists of convex sets of X. A set $X \subset E$ is called c-compact if every convex filter on X has at least one point of accumulation in X. Among the properties of this notion I only mention that for a non-archimedean valued field the following properties are equivalent:

(i) K is c-compact,

(ii) the ring $\{a \in K \mid |a| \leqslant 1\}$ is c-compact,

(iii) K is spherically complete.

2.2. The chapter concerning the separation of convex sets by hyperplanes can be generalized for locally K-convex spaces. In the absence of an ordering of K, it is not possible to define half-spaces, determined by a hyperplane, in the usual way. The following method proved to be useful [*16*].

For $x, y \in E$, the convex hull $C(\{x, y\})$ of the set $\{x, y\}$ is the set of the points $\lambda x + (1-\lambda) y$ where $\lambda \in K$, $|\lambda| \leqslant 1$. Let H be the hyperplane $F(x) = \alpha$, $\alpha \in K$. Separation of points is defined in the following way:

1. The points $x, y \in E$, $x \notin H$, $y \notin H$, are said to be separated by H if $C(\{x, y\}) \cap H \neq \emptyset$.

2. The points $x, y \in E$, $x \notin H$, $y \notin H$, are said to lie on the same side of H if $C(\{x, y\}) \cap H = \emptyset$. We note $x \sim y$ is x, y lie on the same side of H.

It is easily proved that $\sim$ is an equivalence relation. The classes which are determined by this relation are called the "sides of H". In general a hyperplane H has many sides. With the aid of these definitions and with certain restrictions on K it is possible to prove theorems about separation of K-convex sets by hyperplanes which are analogous to the well-known theorems in spaces over R. Lying in a halfspace has then the meaning "being contained in an equivalence class". In particular, the theorem of Hahn-Banach in its geometrical form (see [*13*]) is valid if K is spherically complete.

A generalization of the theory of supporting hyperplanes presents difficulties, owing to the fact that, as I already mentioned, the boundary of convex sets is in general empty. Therefore, it seems to be necessary to define a boundary with the aid of other, non-topological methods. Such a definition has been given in [*16*], but it is not quite satisfactory. A definition of extremal points, for instance, met with difficulties and thus it is unknown whether it is possible to prove an analogue of the theorem of Krein-Milman. On the other hand, this notion of boundary turned out to be useful in proving a maximal principle for a certain class of functions which in a certain sense is analogous to the maximal principle in complex function theory [*17*]. In view of these difficulties a definition of refinements of the notion of convexity in analogy to the theory of linear spaces over the reals (uniform convexity, strict convexity) seems for the moment not possible.

§ 3. Applications

I give a brief survey of further researches and of applications of the theory.

1. Linear operators were studied, in particular completely continuous operators. The theory of Riesz-Schauder in normed spaces over local fields was developed in analogy to the theory for spaces over R (Monna, Serre, Gruson). Van Tiel has put forward a theory of these operators in locally K-convex spaces. A theory of projections was given earlier on in connection with the notion of orthogonality which I mentioned before [*18*].

2. Integration. A theory of integration for functions defined on a locally compact topological space X with values in K is developed in close analogy to the method of Bourbaki [*19*]. The starting point is the space $C_c(X)$. A measure is defined as a linear map $C_c(X) \to K$. The space L^1 is then defined as the closure of $C_c(X)$ with respect to a new norm. Among the results I mention that there is a maximal null set; in a certain sense the functions of L^1 are continuous almost everywhere. In the same paper the existence of a Haar-measure with values in K on a locally compact zero dimensional topological group G is treated. The situation is more complicated than for real measures. A non-trivial Haar measure only exists under a condition on G, depending on the characteristic of the residue class field of K [*20*].

3. Various other problems were studied. I mention fix-point theorems [*21*], summation methods [*22*], an extensive study of Robert [*23*] concerning normed spaces over fields with a trivial valuation where he gives applications on asymptotics and a study of VAN DER PUT on algebras of continuous functions with applications to a spectral analysis of the elements of the algebra.

Notes and References

[1a] MONNA, A. F.: Over niet-archimedische lineaire ruimten. Ned. Akad. v. Wetensch. **52**, 308–321 (1943).

[1b] MONNA, A. F.: Over een lineaire *P*-adische ruimte. Ned. Akad. v. Wetensch. **52**, 74–82 (1943).

[2] For the older literature about normed spaces see the references given in the article by VAN TIEL [14].

[3] This notion was introduced by A. W. INGLETON in his paper The Hahn-Banach theorem for non archimedean valued fields, Proc. Cambridge Phil. Soc. **48**, 41–45 (1952). See the remarks on the binary intersection property in the paper by NACHBIN [7].

[4] CARPENTIER, J.-P.: Semi-normes et ensembles convexes dans un espace vectoriel sur un corps ultramétrique. Séminaire Choquet, **4** (1964/1965).

[5] OSTROWSKI, A.: Untersuchungen zur arithmetischen Theorie der Körper. Math. Z. **39**, 269–404 (1935).

[6] KAPLANSKY, I.: Maximal fields with valuations. Duke Math. J. **9**, 303–321 (1942).

[7] LAZARD, M.: Les zéros d'une fonction analytique d'une variable sur un corps valué complet. Inst. des Hautes Etudes Scientifiques **14** (1962).

[8] NACHBIN, L.: A theorem of the Hahn-Banach type for linear transformations. Trans. Am. Math. Soc. **68**, 28–40 (1950). For a review of the problems on extending and lifting of linear maps see also:
(i) KÖTHE, G.: Fortsetzung linearer Abbildungen lokalkonvexer Räume. Jahresbericht der Deutschen Math. Ver. **68**, 193–204 (1966).
(ii) NACHBIN, L.: Some problems in extending and lifting continuous linear transformations. Proc. Intern. Symp. Linear spaces (Jerusalem 1960). 340–351, Jerusalem: Ac. Press 1961.
(iii) for a general treatment of these questions in terms of categories:
SEMADENI, Z.: Projectivity, injectivity and duality. Warszawa: Rozprawy Matematyczne 1963.

[9] See the references given in VAN TIEL [14].

[10] MONNA, A. F.: Sur les espaces non-archimédiens, I, II. Proc. Kon. Ned. Akad. v. Wetensch. **A.59**, 475–489 (1956).
FLEISCHER, I.: Sur les espaces normés non-archimédiens. Proc. Kon. Ned. Akad. v. Wetensch. **A 57**, 165–168 (1954).
MONNA, A. F., and T. A. SPRINGER: Sur la structure des espaces de Banach non-archimédiens. Proc. Kon. Ned. Akad. v. Wetensch. **A 68**, 602–614 (1965).

[11a] MONNA, A. F.: Ensembles convexes dans les espaces vectoriels sur un corps valué. Proc. Kon. Ned. Akad. v. Wetensch. **A 61**, 528–539 (1958).

[11b] MONNA, A. F.: Espaces localement convexes sur un corps valué. Proc. Kon. Ned. Akad. v. Wetensch. **A 62**, 391–405 (1959).

[12] SPRINGER, T. A.: Une notion de compacité dans la théorie des espaces vectoriels topologiques. Proc. Kon. Ned. Akad. v. Wetensch. **A 68**, 182–189 (1965).

[13] See for instance: BOURBAKI, N.: Espaces vectoriels topologiques, Act. Sci. et Ind. 1189 (1953).

[14] TIEL, J. VAN: Espaces localement K-convexes, Proc. Kon. Ned. Akad. v. Wetensch. **A 68**, 249–258, 259–272, 273–289 (1965).
TIEL, J. VAN: Ensembles pseudo-polaires dans les espaces localement K-convexes. Proc. Kon. Ned. Akad. v. Wetensch. **A 69**, 369–373 (1966).

[15] KLEE, V. L.: Convex sets in linear spaces I, II, Duke Math. Journal **18**, 443–466, 875–883 (1951).
I mention also a characterization of reflexivity in terms of problems of shortest distance; see BOR-LUH LIN, Distance sets in normed vector spaces, Nieuw Archief voor Wiskunde **14**, 23–30 (1966).

[16] MONNA, A. F.: Séparation d'ensembles convexes dans un espace linéaire topologique sur un corps valué. Proc. Kon. Ned. Akad. v. Wetensch. **A 67**, 399–408, 409–421 (1964).
For further results and some improvements see CARPENTIER [4].

[17] MONNA, A. F.: Sur un principe de maximum en analyse P-adique. Proc. Kon. Ned. Akad. v. Wetensch. **A 69**, 213–222 (1966).
In mention here the work of GRAUERT und REMMERT, Nichtarchimedische Funktionentheorie, Festschr. Gedächtnisfeier K. Weierstrass, 393–476, Köln: Westdeutscher Verlag 1966.

[18] For projections see the older literature. For linear operators in normed spaces see: MONNA, A. F.: Lineaire functionaalvergelijkingen in niet-archimedische Banach-ruimtes. Ned. Akad. v. Wetensch. **A 52**, 654–661 (1943). See also: SERRE, J.-P.: Endomorphisms complètement continus des espaces de Banach p-adiques. Inst. H. E. Sci, no. 12 (1962); GRUSON, L.: Théorie de Fredholm p-adique. Bull. Soc. Math. France **94**, 67–95 (1966).
A general theory, based on the theory of locally K-convex spaces is given by VAN TIEL, J.: Une note sur les applications complètement continues. Proc. Kon. Ned. Akad. v. Wetensch. **A 68**, 772–776 (1965).

[19] MONNA, A. F., and SPRINGER, T. A.: Intégration non-archimédienne. I, II, Proc. Kon. Ned. Akad. v. Wetensch. **A 66**, 634–642, 643–653 (1963).
See also: TOMAS, F.: Integracion p-adica. Boletin de la Sociedad Matematica Mexicana, 1–38 (1962).
See also BRUHAT, F.: Intégration p-adique. Sém. Bourbaki, 229 (1961–62).

[20] See also A. C. H. van ROOIJ: Proc. Kon. Ned. Akad. v. Wetensch. **A LXX**. 220–228 (1967), where results on the existence of invariant means are obtained. The only property of K relevant to the existence of a means is the characteristic of the residue class field.

[21] MONNA, A. F.: Note sur les points fixes. Proc. Kon. Ned. Akad. v. Wetensch. **A 67**, 588–593 (1964).

[22] MONNA, A. F. : Sur le théorème de Banach-Steinhaus. Proc. Kon. Ned. Akad. v. Wetensch. **A 66**, 121–131 (1963).
RANGACHARI, M. S., and V. K. SRINIVASAN: Matrix transformations in non-archimedean valued fields. Proc. Kon. Ned. Akad. v. Wetensch. **A 67**, 422–429 (1964).
SRINIVASAN, V. K.: On certain summation processes in the p-adic field. Proc. Kon. Ned. Akad. v. Wetensch. **A 68**, 319–325 (1965).

[23] ROBERT, P.: On some non-archimedean normed linear spaces (to appear).

Modèles minimaux des espaces principaux homogènes sur les courbes elliptiques

A. Néron

Le théorème d'existence d'un modèle $\mathfrak{p}$-simple $\mathfrak{p}$-minimal pour une courbe elliptique n'est démontré dans [6], chap. III que dans le cas où cette courbe possède un point rationnel sur le corps de base (local ou global) k, c'est-à-dire dans le cas où l'on peut munir cette courbe d'une structure de variété abélienne A, définie sur k. Je me propose d'exposer ici, tout au moins dans ses grandes lignes, une démonstration d'un théorème d'existence analogue, valable pour une courbe elliptique quelconque définie sur k, mais ne possédant pas nécessairement un point rationnel sur k — ou, si l'on préfère, valable pour un espace principal homogène sur A, défini sur k.

1. Rappel sur les espaces principaux homogènes

Soit G une variété de groupe définie sur un corps k quelconque. On appelle *espace principal homogène* sur G, défini sur k, une variété V, définie sur k, sur laquelle G opère au moyen d'un k-morphisme $\lambda: G\times V\to V$; on suppose en outre que G opère simplement et transitivement, i.e. que, pour x et $y\in V$, il existe un et un seul $g\in G$ tel qu'on ait $y=\lambda(g, x)$, et tel que l'application $\mu: V\times V\to G$ obtenue en posant $\mu(x, y)=g$ soit un morphisme. Pour $a\in V$, l'application $\lambda_a: G\to V$ obtenue en posant $\lambda_a(g)=\lambda(g, a)$ est donc un isomorphisme (pour la structure de variété algébrique) sur le corps $k(a)$.

Deux espaces principaux homogènes V et V' sur G, définis sur k, sont dits *k-isomorphes* s'il existe un k-isomorphisme $V\to V'$ (pour la structure de variété algébrique) compatible avec l'opération de G. Supposons G commutatif, et soit K une extension de k. L'ensemble des classes, pour la relation de k-isomorphisme, d'espaces principaux homogènes sur G possédant un point rationnel sur K est naturellement muni d'une structure de groupe [*10*]. Ce groupe est appelé le *groupe de Weil-Châtelet* de G relativement à l'extension K/k. Si l'extension K/k est galoisienne, de groupe de Galois Γ, on sait que ce groupe est isomorphe au premier groupe de cohomologie galoisienne $H^1(K/k, G)$ [*1*], [*2*], [*5*]. Plus précisément, cet isomorphisme peut être construit comme suit: V étant un espace principal homogène sur G, défini sur k, possèdant un point v_0 rationnel sur k, on lui fait correspondre la classe de cohomologie représentée par le cocycle $\sigma \mapsto c_\sigma$ obtenu en posant

$$c_\sigma = \mu(v_0^\sigma, v_0).$$

2. Rappel sur les modèles minimaux

Soit k un corps muni d'une valuation discrète v, de corps résiduel k^0 parfait. Comme dans [*6*], on désigne par R l'anneau des entiers de k, par $\mathfrak{p}$ son idéal maximal, et par t une uniformisante, i.e. un générateur de $\mathfrak{p}$, par p la caractéristique de k^0.

Soit W une $\mathfrak{p}$-variété définie sur k (par exemple une variété projective définie sur k). On sait définir le cycle $W^0 = \varrho(W)$ réduit de $W \pmod{\mathfrak{p}}$ et, à tout point w^0 du support $\rho_e(W) = \operatorname{supp} W^0$, faire correspondre son anneau local $\mathfrak{o}(w^0, W)$. Il correspond canoniquement à la variété W un schéma $\mathbf{S}_R(W)$ sur R au sens de GROTHENDIECK [*3*], admettant pour anneaux locaux, d'une part ceux des points de la «fibre générique» W, et d'autre part ceux des points de la «fibre spéciale» $\operatorname{supp} W^0$. On dit que W est $\mathfrak{p}$-*simple* si le schéma $\mathbf{S}_R(W)$ est régulier, i.e. si tous ses anneaux locaux sont réguliers.

On dit qu'une application rationnelle $\varphi: W \to W'$ est $\mathfrak{p}$-*morphique* en un point $w^0 \in \operatorname{supp} W^0$ si les coordonnées de φ relativement à un ouvert affine convenable de W′ appartiennent à l'anneau local $\mathfrak{o}(w^0, W)$. On dit que φ est un *R-morphisme* si φ est un k-morphisme et si, de plus, φ est $\mathfrak{p}$-morphique en tout point de $\operatorname{supp} W^0$. Il revient au même de dire que φ provient d'un morphisme $\mathbf{S}_R(W) \to \mathbf{S}_R(W')$ pour la structure de schéma sur R.

On dit qu'une $\mathfrak{p}$-variété $\mathfrak{p}$-simple W_* est $\mathfrak{p}$-*minimale* si tout k-isomorphisme de la forme $\bar{W} \to W_*$, où $\bar{W}$ est une $\mathfrak{p}$-variété $\mathfrak{p}$-simple, est un R-morphisme. On appelle *k-modèle* $\mathfrak{p}$-*simple* $\mathfrak{p}$-*minimal* d'une $\mathfrak{p}$-variété W un couple (W_*, α) composé d'une $\mathfrak{p}$-variété $\mathfrak{p}$-simple $\mathfrak{p}$-minimale W_*, et d'un k-isomorphisme $W \to W_*$. La variété W étant donnée, un tel modèle, s'il existe, est nécessairement unique (i.e. $\mathbf{S}_R(W_*)$ est nécessairement unique) à un R-isomorphisme près. Le schéma sur le corps résiduel k^0 déduit du précédent par réduction (mod $\mathfrak{p}$) est également unique à un k^0-isomorphisme près. Ceci entraîne en particulier l'unicité du nombre des composantes du cycle réduit W^0, de leurs coefficients respectifs dans ce cycle, de leurs multiplicités d'intersection deux à deux, etc.

Le résultat essentiel de [6], III affirme que toute courbe elliptique A définie sur k, et possédant un point rationnel sur k, admet un k-modèle $\mathfrak{p}$-simple $\mathfrak{p}$-minimal A_*. Il en résulte aussitôt que l'ensemble $\mathscr{S}(A_*^0)$ (noté également $\mathscr{G}_\mathfrak{p}(A)$) des points simples sur A_*^0 est canoniquement muni d'une structure de *groupe algébrique*, déduite de celle de A par réduction (mod $\mathfrak{p}$). La méthode de démonstration utilisée fournit en outre une classification complète des différents types de réduction possibles pour A_* ([6], III, 17).

Mais il est possible de simplifier ce type de réduction en agrandissant k; de ce point de vue, deux cas sont à distinguer:

(*) – *L'invariant $j = j(A)$ est un entier* – Alors il existe une extension algébrique k_1 de k, de degré fini ν divisant 24, tel que tout modèle $\mathfrak{p}_1$-simple $\mathfrak{p}_1$-minimal A_{1*} de A sur k_1 (où l'on note $\mathfrak{p}_1$ l'idéal de la valuation discrète v_1 de k_1 prolongeant v) ait une *bonne réduction*, i.e. tel que $A_{1*}^0 = \varrho(A_{1*}^0)$ soit une courbe elliptique.

(**) – *L'invariant $j = j(A)$ n'est pas un entier* – Alors il existe une extension algébrique k_1 de k, de degré $\nu = 1$ ou 2, telle que, en désignant, comme ci-dessus, par A_{1*} un modèle $\mathfrak{p}_1$-simple $\mathfrak{p}_1$-minimal de A sur k_1, la composante de l'origine du groupe réduit $\mathscr{S}(A_{1*}^0)$ soit isomorphe au *groupe multiplicatif*. Si l'on pose $\nu\, v(j) = -m$ (m entier naturel), le cycle A_{1*}^0 admet m composantes simples; pour $m=1$, l'unique composante est une courbe de genre 0 admettant un point double à tangentes distinctes; pour $m>1$, les m composantes sont des courbes de genre 0 sans point double, et sont connectées à la façon des côtés d'un polygone.

Un cycle de cette espèce sera dit *polygonal* de m côtés.

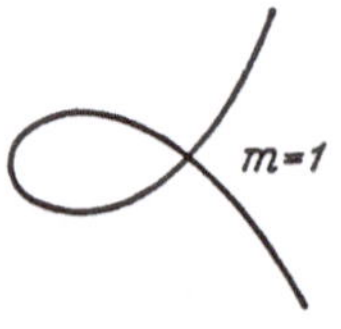

Fig. 1

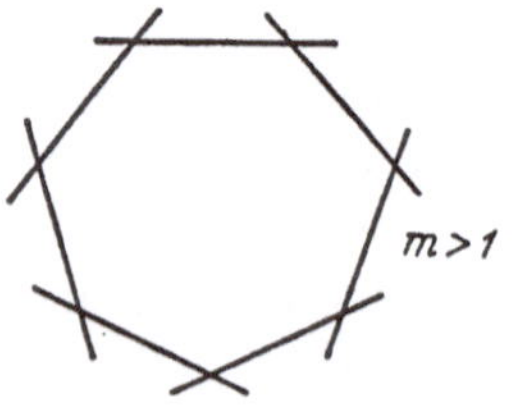

Fig. 2

Pour k et A fixés, on peut, dans tous les cas, déterminer la valeur minimum possible v_0 de l'entier v. On peut supposer que A est une courbe elliptique $\mathfrak{p}$-*standard* au sens de [*6*], III.

Les conditions (a) ou (c_i) $(i \neq 5)$ de [*6*], III, prop. 4 entraînent (*), et on a respectivement $v_0 = 1$ dans le cas (a), $v_0 = 2$ dans le cas $(c4)$, $v_0 = 3$ dans les cas $(c3)$ et $(c6)$, $v_0 = 4$ dans les cas $(c2)$ et $(c7)$, $v_0 = 6$ dans les cas $(c1)$ et $(c8)$.

La condition (b) entraîne (**), avec $v_0 = 1$.

Si la caractéristique p du corps résiduel k^0 est $\neq 2$, la condition $(c5)$ entraine (**), avec $v_0 = 2$. Il n'en est plus de même si on a $p = 2$, mais le cas $(c5)$ se ramène alors à l'un de ceux précédemment énumérés, après remplacement de k par une extension quadratique convenable. En définitive, le cas $(c5)$ se subdivise en deux sous-cas non vides, respectivement inclus dans (*) et (**), qu'on désignera par $(c5^*)$ et $(c5^{**})$, ce dernier impliquant $p = 2$.

3. Enoncé du théorème d'existence

Théorème 1 – *Soit V un espace principal homogène sur A, défini sur k. Alors il existe un k-modèle $\mathfrak{p}$-simple $\mathfrak{p}$-minimal (W, γ) de V.*

Observons que le problème n'est pas essentiellement modifié lorsqu'on remplace k par une extension algebrique, éventuellement infinie, *non ramifiée*, k' de k. Si (W', γ') est une solution du problème pour k', on peut en effet en déduire une solution du problème pour k par descente non ramifiée du corps de base, et les types de réduction respectivement obtenus pour W' et pour W sont les mêmes. On peut donc, en particulier, se ramener au cas où *le corps résiduel k^0 est algébriquement clos.*

Désignons par $r = r(V)$ *l'indice séparable* de V au sens de [*5*], i.e. le p.g.c.d. des degrés des extensions séparables K' de k telles que V possède un point rationnel sur K'. On peut, en se limitant au cas non trivial $r > 1$, compléter l'énoncé du th. 1 en précisant comme suit le type de réduction obtenu pour W.

Théorème 1' – *Les notations étant celles du th. 1, supposons k^0 algébriquement clos, et* $\mathrm{r} > 1$.

Dans le cas (*), le cycle réduit W^0 de W est de la *forme* $W^0 = r E^0$, *où* E^0 *est une courbe elliptique.*

Dans le cas (**), et en posant $v(j) = -m$, on a $W^0 = r P^0$, où P^0 est un cycle polygonal de m côtés.

Enonçons encore le résultat relatif au cas d'un *corps global L* (un tel corps L est muni d'une famille de valuations discrètes, comme il est précisé dans [6], I, 26, et on note $\mathfrak{S}=\mathfrak{S}(L)$ l'ensemble des $\mathfrak{p}$ correspondants).

Théorème 2 (cas global) – *A étant supposée définie sur un corps global L, soit V un espace principal homogène sur A, défini sur L. Alors il existe un L-modèle* (W, γ) *de V qui est* $\mathfrak{p}$*-simple* $\mathfrak{p}$*-minimal pour tout* $\mathfrak{p}\in\mathfrak{S}(L)$.

4. Ensemble des valeurs prises par l'entier r

On peut chercher à déterminer, pour k donné, et pour A donnée $\mathfrak{p}$-standard, l'ensemble des valeurs possibles de l'entier $r=r(V)$ intervenant dans le th. 1'. Il y a lieu de distinguer à nouveau les différents cas de la classification de [6], III, 7.

Dans le cas (a), et si on a $p=0$, ou si on a $p\neq 0$, et si l'invariant de Hasse h de la courbe réduite A^0 est non nul, on trouve, quel que soit A, que r peut prendre toutes les valeurs entières. J'ignore s'il en est encore ainsi lorsqu'on a $p\neq 0$ et $h=0$; on peut toutefois affirmer que r prend alors toutes les valeurs entières premières à p.

Dans le cas (b), on trouve, quel que soit A, que r peut prendre toutes les valeurs entières.

Dans les autres cas, une réponse partielle est fournie par le théorème suivant (dans lequel ons uppose toujours que k^0 est algébriquement clos).

Théorème 3

(i) – *Si A vérifie* (c5**), *on a* $H^1(k, A)=0$.

(ii) – *Supposons que A vérifie* (c_i) $(i\neq 5)$ *ou* (c5*); *si on a* $p=0$, *ou si on a* $p\neq 0$, *et si l'invariant de Hasse* h_1 *de* A^0_{1*} *n'est pas nul, on a encore* $H^1(k, A)=0$.

(Dans cet énoncé, A^0_{1*} désigne la courbe elliptique réduite du modèle minimal A_{1*} de A sur k_1 introduit au n^0 2; le symbole $H^1(k, A)$ est mis, selon l'usage, pour $H^1(k_s/k, A)$ où k_s est la clôture séparable de k.)

Il revient au même de dire que, dans l'une ou l'autre des hypothèses de ce théorème, l'entier r ne peut prendre que la valeur 1. J'ignore si le résultat est encore valable dans (ii) lorsqu'on a $p\neq 0$[1] et $h_1=0$ (s'il en était ainsi, le groupe $H^1(k, A)$ serait toujours nul dans l'un quelconque des cas (c_i), i.e. toutes les fois que la composante de l'origine du groupe réduit A^0 est isomorphe au groupe additif); on peut toutefois montrer que, dans ce cas, r est nécessairement une puissance de p.

[1] Il en, est bien ainsi, comme il résulte d'une démonstration très simple que vient de me communiquer M. RAYNAUD.

Les démonstrations des résultats de ce n° peuvent s'obtenir directement en utilisant les calculs explicites de [6], III; nous ne les reproduirons pas ici (bien que l'assertion (i) du th. 3 soit utilisée au n° suivant, et intervienne dans la démonstration du th. 1). Une bonne partie de ces résultats, comprenant la détermination de l'ensemble des valeurs *premières à p* prises, pour k et A donnés, par l'entier r, pourrait se déduire des résultats de Ogg [7] et Šafarevič [9] (voir aussi [8]). Notons que lorsque $r=r(V)$ est premier à p, il est encore égal à la *période* de V au sens de [5], i.e. à l'ordre de l'élément du groupe de Weil-Châtelet représenté par V (cf. [7], th. 1, coroll. 3).

5. Démonstration du théorème 1: remarques et conventions préliminaires

On peut, comme on a vu, supposer que le corps résiduel k° est algébriquement clos. On désigne par K une extension algébrique galosienne de degré fini de k, contenant le corps k_1 introduit au n° 2, et telle que V possède un point rationnel sur K. L'unique valuation de K prolongeant v est notée w. On désigne respectivement par S et $\mathfrak{q}$ l'anneau et l'idéal de valuation correspondants, et par u une uniformisante de S. On note Γ le groupe de Galois $\mathrm{Gal}(K/k)$ de l'extension K/k. On désigne par $c:\sigma\longmapsto c_\sigma$ (σ parcourant Γ) un cocycle appartenant à la classe $\kappa\in H^1(K/k, A)$ associée à V (cf. n° 1). On peut, comme on sait, supposer c à valeurs dans le groupe des points d'ordre fini de A.

Introduisons un K-isomorphisme $\varphi: A\to B$ de A sur un modèle $\mathfrak{q}$-simple $\mathfrak{q}$-minimal B de A. Comme on a supposé $K\supset k_1$, la variété B admet une bonne réduction, ou une réduction de type polygonal, suivant qu'on est dans le cas (*) ou (**). Pour $\sigma\in\Gamma$, notons $\varphi^\sigma: A\to B^\sigma$ le K-isomorphisme conjugué de φ par σ, ayant pour image la courbe B^σ conjuguée de B. Comme k^0 est algébriquement clos, le cycle réduit $(B^\sigma)^0$ de B^σ coïncide avec B^0. Considérons les morphismes $\lambda: A\times V\to V$, $\mu: V\times V\to A$ définissant la structure d'espace principal homogène de V. Il existe un point $v_0\in V$, rationnel sur K, tel que $c_\sigma=\mu(v_0^\sigma, v_0)$. Soit x un point générique de de A sur K. Introduisons le K-isomorphisme $\eta=\lambda_{v_0}: A\to V$ obtenu en posant $\eta(x)=\lambda(x, v_0)$ et considérons les points génériques $y=\varphi(x)$ et $z=\eta(x)$ de B et de V respectivement sur K. Le corps $k(z)$ est isomorphe au corps des invariants de $K(x)=K(y)=K(z)$ par les k-automorphismes de la forme

$$\sigma^*: f(x)\longmapsto f^\sigma(x+c_\sigma),$$

avec $\sigma\in\Gamma$; l'application $\sigma\longmapsto\sigma^*$ est un isomorphisme du groupe de Galois

$\Gamma = \mathrm{Gal}(K/k)$ sur le groupe de Galois $\Gamma^* = \mathrm{Gal}(K(x)/k(z))$. Le point x^{σ^*} de A (resp. le point y^{σ^*} de B^σ) transformé de x (resp. de $y = \varphi(x)$) par σ^* est défini par $x^{\sigma^*} = x + c_\sigma$ (resp. par $y^{\sigma^*} = \varphi^\sigma(x + c_\sigma)$, c'est-à-dire par $y^{\sigma^*} = \psi_\sigma(y)$, en désignant par ψ_σ le K-isomorphisme $B \to B^\sigma$ défini par $\psi_\sigma = \varphi^\sigma \circ T_{c_\sigma} \circ \varphi^{-1}$, où T_{c_σ} est la translation $x \longmapsto x + c_\sigma$ sur A); on a aussi $\psi_\sigma = \xi_\sigma \circ T_{d_\sigma}$, où ξ_σ est le K-isomorphisme $\varphi^\sigma \circ \varphi^{-1}$ de B sur B^σ, et où T_{d_σ} est la translation $y \longmapsto y + d_\sigma$ sur B, définie par le point $d_\sigma = \varphi(c_\sigma)$.

D'après les propriétés des modèles minimaux, chacun des K-isomorphismes ξ_σ, T_{d_σ}, ψ_σ est $\mathfrak{q}$-isomorphique en tout point $b^0 \in \operatorname{supp} B^0$. Le point $\psi_\sigma^0(b^0)$ de $\operatorname{supp} B^0$ (qu'on pourrait encore désigner par $(b^0)^{\sigma^*}$) est dit *conjugué de b^0 par σ^**; il coïncide avec b^0 lorsque σ (i.e. lorsque σ^*) est l'identité. La famille des points $\{\psi_\sigma^0(b^0)\}_{\sigma \in \Gamma}$ est appelée *l'orbite* de b^0. Les points c_σ étant rationnels sur K, leurs points réduits c_σ^0 appartiennent au groupe $\mathscr{S}(B^0)$. Il en résulte que si $b^0 \in \mathscr{S}(B^0)$, il en est de même de ses conjugués par les éléments de σ^*.

Mentionnons encore une réduction du problème particulière au cas (**) où j est non entier. La courbe A, supposée $\mathfrak{p}$-standard, vérifie l'une des conditions (b) ou ($c5$**). Le cas ($c5$**) étant trivial, en vertu de l'assertion (i) du th. 3, il suffit d'examiner le cas (b). Or, dans ce dernier cas, pour $b^0 \in \mathscr{S}(B^0)$ et pour $\sigma \in \Gamma$, on trouve que les points b^0 et $\psi_\sigma^0(b^0)$ appartiennent à une même composante de $\mathscr{S}(B^0)$. Les éléments tels que le point réduit d_σ^0 de $d_\sigma = \varphi(c_\sigma)$ appartienne à la composante de l'origine de $\mathscr{S}(B^0)$ forment un sous-groupe distingué Γ_0 de Γ. On peut montrer qu'il existe un cocycle équivalent à c s'annulant sur Γ_0. La classe κ de c s'identifie donc à un élément du groupe $H^1(K_0/k, A)$, où K_0 est le sous-corps de K formé des éléments invariants par Γ_0. On peut donc remplacer K par K_0, et on se ramène ainsi au cas où Γ_0 est réduit à l'identité. Dans ces conditions, pour $b^0 \in \mathscr{S}(B^0)$ donné, les points $\psi_\sigma^0(b^0)$ $(\sigma \in \Gamma)$ formant l'orbite de b^0 sont distincts, et appartiennent à des composantes *distinctes* de $\mathscr{S}(B^0)$. De plus, si s^0 est un sommet de B^0, les points $\psi_\sigma^0(b^0)$ formant l'orbite de s^0 sont des sommets *distincts* de B^0.

6. Construction du modèle (W, γ)

Introduisons le diviseur

$$X = \Sigma_{\sigma \in \Gamma}(c_\sigma)$$

sur A, et ses images respectives $Y = \varphi(X)$ sur B et $Z = \eta(X)$ sur V. On a $Z = \Sigma_\sigma(v_\sigma^0)$, donc Z est rationnel sur k.

Notons E le k-espace vectoriel composé des fonctions f sur A, définies sur K, invariantes par σ^*, et telles que $\operatorname{div}(f) \geqslant -3X$. Notons M le R-module composé des éléments f de E tels que $f \circ \varphi$ soit génériquement $\mathfrak{p}$-morphique sur chacune des composantes de B^0, ou, ce qui revient au même, tels qu'on ait $\operatorname{div}_q(f \circ \varphi) \geqslant 0$, en désignant par div_q le q-diviseur au sens de [6], I, 12.

Le module M est libre de type fini. Soit $\mathscr{B} = \{f_0, \ldots, f_n\}$ une base de M. Désignant toujours par x un point générique de A sur K, notons w_0 le point de l'espace projectif $\mathbf{P}_n$ ayant pour système de coordonnées homogènes les $w_{0_j} = f_j(x)$; soit W_0 le lieu de w_0 sur k, et soit $\alpha_0: A \to W_0$ l'application rationnelle définie sur k, telle que $\alpha_0(x) = w_0$.

Montrons que l'application rationnelle $\gamma_0 = \alpha_0 \circ \eta^{-1}: V \to W_0$ est un k-isomorphisme. En effet, quelle que soit $f \in E$, il existe un entier l tel que $f t^l \in M$; donc $\mathscr{B}$ est une base de E sur k. Or E s'identifie à l'espace vectoriel $\mathscr{L}(3Z)$ composé des fonctions h sur V, définies sur k, telles que $\operatorname{div}(h) \geqslant -3Z$. Donc γ_0 est un plongement projectif associé au système linéaire complet déterminé par le diviseur $3Z$. Comme ce dernier est ample, γ_0 est bien un k-isomorphisme.

Désignons par (W, θ) un modèle $\mathfrak{p}$-normalisé canonique de W_0: un tel couple est composé d'une variété $\mathfrak{p}$-normale W et d'un k-isomorphisme $\theta: W_0 \to W$ tel que $\theta^{-1}: W \to W_0$ soit un R-morphisme, et que l'ensemble des valeurs de θ en tout point du support de $W_0^0 = \varrho_e(W_0)$ soit fini. Posons $\alpha = \theta \circ \alpha_0$, $\beta_0 = \varphi \circ \alpha_0^{-1}$, $\beta = \alpha \circ \varphi^{-1} = \theta \circ \beta_0$, $\gamma = \alpha \circ \eta^{-1} = \theta \circ \gamma_0$. On a ainsi le diagramme commutatif suivant de K-isomorphismes

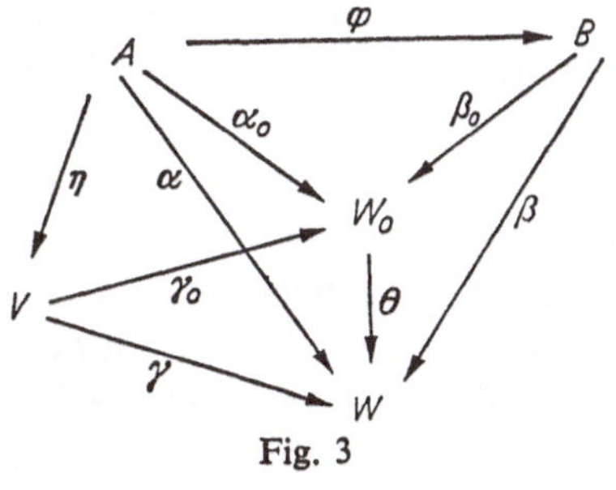

Fig. 3

7. Justification de la construction précédente; suite et fin de la démonstration du théorème 1

Il est nécessaire ici de rappeler les conventions faites dans [6] quant au choix de domaines universels pour les corps k^0 et k: on a d'abord choisi un domaine universel $\mathfrak{k}^0$ pour k^0; on a introduit l'extension $\mathfrak{k}$ de k obtenue «en relevant $\mathfrak{k}^0$», et construite au moyen des séries formelles ou des vecteurs de Witt à coefficients dans $\mathfrak{k}^0$; rappelons que $\mathfrak{k}$ est complet pour une valuation canonique $\bar{v}$ prolongeant v; on a choisi ensuite un domaine universel arbitraire Ω pour k.

Nous pouvons supposer que K est un sous-corps de Ω. Le corps $\mathfrak{K}$, analogue à $\mathfrak{k}$, construit à partir de K et w, s'identifie au composé $K\mathfrak{k}$; sa valuation canonique $\bar{w}$ prolonge w, et c'est aussi l'unique valuation de $\mathfrak{K}$ prolongeant $\bar{v}$; par abus de langage, on confondra dans la suite $\bar{v}$ avec v, et $\bar{w}$ avec w. On désignera par $\mathfrak{R}$ (resp. $\mathfrak{S}$) l'anneau des entiers de $\mathfrak{k}$ (resp. $\mathfrak{K}$). L'extension $\mathfrak{K}/\mathfrak{k}$ est galoisienne, et on a un isomorphisme canonique $\Gamma=\mathrm{Gal}(K/k)\to\mathrm{Gal}(\mathfrak{K}/\mathfrak{k})$ (resp. $\Gamma^*=\mathrm{Gal}(K(x)/k(z))\to\mathrm{Gal}(\mathfrak{K}(x)/\mathfrak{k}(z))$). L'image par cet isomorphisme d'un élément $\sigma\in\Gamma$ (resp. $\sigma^*\in\Gamma^*$) est un automorphisme prolongeant σ (resp. σ^*) que, par abus de langage, on notera encore σ (resp. σ^*).

On désignera par $\mathfrak{K}$ le $\mathfrak{k}$-espace vectoriel composé des fonctions f sur A, définies sur $\mathfrak{K}$, invariantes par Γ^*, et telles que $\mathrm{div}(f)\geqslant -3X$. Compte tenu de [*11*], IX, 4, th. 8, on a $\mathfrak{E}\simeq E\otimes_k\mathfrak{k}$. De même, on désignera par $\mathfrak{M}$ le $\mathfrak{R}$-module composé des $f\in E$ telles qu'on ait $\mathrm{div}_q(f\circ\varphi)\geqslant 0$. Compte tenu de [*6*], II, prop. 8, on a $\mathfrak{M}\simeq M\otimes_R\mathfrak{R}$.

Nous allons maintenant montrer que $\beta: B\to W$ est un S-morphisme et que, de plus, deux points de $\mathrm{supp}B^0$ ont même image par β si et si seulement ils appartiennent à une même orbite.

Considérons trois points b_1^0, b_2^0, b_3^0 de la composante de l'origine de $\mathscr{S}(B^0)$, tels que $b_1^0+b_2^0+b_3^0=0$. D'après le lemme de Hensel (sous la forme donnée dans [*6*], I, 10, prop. 5) on peut les relever respectivement à des points b_1, b_2, b_3 de B, rationnels sur $\mathfrak{K}$, tels que $b_1+b_2+b_3=0$. Posons $a_i=\varphi^{-1}(b_i)$ $(i=1, 2, 3)$. Il existe une fonction g sur B, définie sur K, telle que $\mathrm{div}(g)=(b_1)+(b_2)+(b_3)-3\,(0)$, et telle que $\mathrm{div}(g)=0$ (d'après [*6*], II, 6, prop. 5). La fonction $f=\prod_{\sigma^*\in\Gamma^*}(g\circ\varphi)^{\sigma^*}$ sur A est invariante par Γ^* et a pour diviseur $X'-3X$, en posant $X'=\Sigma_{\sigma,i}\,(a_i^\sigma-c_\sigma)$. On a donc $f\in M$. D'autre part, le cycle réduit de $Y=\varphi(X)=\Sigma_{\sigma,i}\,(\xi_\sigma^{-1}(b_i^\sigma)-d_\sigma)=\Sigma_{\sigma,i}\,\psi_\sigma^{-1}(b_i^\sigma)$ admet pour composants les points des orbites des b_i^0 $(i=1, 2, 3)$.

Soient maintenant b^0 et b'^0 deux points de $\mathscr{S}(B^0)$ n'appartenant pas à une même orbite. On peut choisir deux triplets (b_1, b_2, b_3) et $(\bar{b}_1, \bar{b}_2, \bar{b}_3)$ comme ci-dessus, de façon que $b_1^0=b^0$, et qu'aucun des points b_2^0, b_3^0, $\bar{b}_1^0$, $\bar{b}_2^0$, $\bar{b}_3^0$ n'appartienne à l'orbite de b^0 ni à celle de b'^0. A partir de ces deux triplets, construisons deux éléments f et $\bar{f}$ de $\mathfrak{M}$ comme ci-dessus, de diviseurs respectifs $X'-3X$ et $\bar{X}'-3X$; posons $Y'=\varphi(X')$ et $\bar{Y}'=\varphi(\bar{X}')$. On a $b^0 \in \varrho_e(\operatorname{supp} Y')$, $b^0 \notin \varrho_e(\operatorname{supp} \bar{Y}')$, $b'^0 \notin \varrho_e(\operatorname{supp} Y')$, et $b'^0 \notin \varrho_e(\operatorname{supp} \bar{Y}')$. Donc ([6], I, 12, prop. 10, coroll.) la fonction $(f/\bar{f})$ est morphique en chacun des points b^0 et b'^0, et s'annule en b^0, mais non en b'^0. Donc β_0 est morphique en chacum des points b^0 et b'^0 et, de plus, sépare ces deux points. Il en résulte que $\beta=\beta_0 \circ \theta: B \to W$ ne prend qu'un nombre fini de valeurs en b^0 et en b'^0. Or B est q-simple, donc q-normale. Donc β est un R-morphisme; de plus, θ applique bijectivement supp W_0^0 sur supp W^0; donc β sépare les deux points b^0 et b'^0. Enfin, deux points d'une même orbite ont même image par β_0, donc aussi par β. On a donc prouvé les assertions énoncées plus haut concernant β.

Nous allons maintenant prouver la $\mathfrak{p}$-simplicité de W. Soit $w^0 \in \operatorname{supp} W^0$; supposons d'abord que w^0 est de la forme $\beta^0(b^0)$, avec $b^0 \in \mathscr{S}(B^0)$. Soit $\bar{w}^0$ un point générique sur k^0 d'une des composantes de W^0. Comme W est $\mathfrak{p}$-normale, $\bar{w}^0$ est $\mathfrak{p}$-simple sur W. De plus, il est de la forme $\beta^0(\bar{b}^0)$, où $\bar{b}^0$ est un point générique sur k^0 de l'une des composantes de B^0. Notons ν le K-morphisme $B \times B \times W \to W$ défini par $\nu(y, y', w)=\lambda(\varphi^1(y'-y), w)$. Il est clair que $\bar{w}^0$ appartient à l'ensemble E^0 des valeurs de ν en $(\bar{b}^0, b^0, w^0)$. Notons que ce dernier point est normal sur $B \times B \times V$. Nous allons montrer que ν est morphique (donc nécessairement de valeur $\bar{w}^0$) en ce point. S'il n'en était pas ainsi, E^0 contiendrait en effet une variété de dimension 1, i.e. une composante de W^0. Soit $\bar{w}_1^0$ un point générique d'une telle composante sur le corps $k^0(b^0, \bar{b}^0)$. Ce point est de la forme $\bar{w}_1^0=\beta^0(\bar{b}_1^0)$, où $\bar{b}_1^0$ est générique sur $k^0(b^0, \bar{b}^0)$ de l'une des composantes de B^0. Introduisons encore un point $\bar{b}_1^0$, générique de l'une quelconque des composantes de B^0 sur le corps $k^0(b^0, \bar{b}^0, \bar{b}_1^0)$, et considérons les points $b_1^0=\bar{b}_1^0+b^0-\bar{b}^0$ et $b_2^0=\bar{b}_2^0-b^0-\bar{b}^0$. Puisque $\bar{w}_1^0$ appartient à l'ensemble des valeurs de ν en $(\bar{b}^0, b^0, w^0)$, il appartient aussi à l'ensemble des valeurs de ν en $(\bar{b}_2^0, b_2^0, w^0)$. Donc w^0 appartient à l'ensemble des valeurs de ν en $(b_2^0, \bar{b}_2^0, \bar{w}_1^0)$. Or $b_2^0, \bar{b}_2^0, \bar{w}_1^0$ sont génériques indépendants sur k^0. Donc le point $(b_2^0, \bar{b}_2^0, \bar{w}_1^0)$ est $\mathfrak{p}$-normal sur $B \times B \times W$, et par suite ν est morphique en ce point, de valeur w^0. Donc ν est morphique en $(b_1^0, \bar{b}_1^0, \bar{w}_1^0)$, et également de valeur w^0. Puisqu'on a $\bar{w}_1^0=\beta^0(\bar{b}_1^0)$, on a aussi nécessairement $w^0=\beta^0(b_1^0)$. Or ceci est absurde, puisque b_1^0 est générique sur $k^0(w^0)$ de l'une des composantes de B^0.

On a donc montré que ν est $\mathfrak{p}$-morphique en $(\bar{b}^0, b^0, w^0)$, de valeur w^0. Si b et $\bar{b}$ sont deux points de B, rationnels sur $\mathfrak{K}$, obtenus en relevant respectivement, de façon arbitraire, les points b^0 et $\bar{b}^0$, la translation $\tau: w \mapsto \nu(\bar{b}, b, w)$ sur W est morphique en w^0, de valeur $\bar{w}^0$. Un raisonnement analogue permet de voir que la translation opposée $-\tau$ est morphique en $\bar{w}^0$, de valeur w^0. Puisque $\bar{w}^0$ est $\mathfrak{p}$-simple sur W, il en est de même de w^0. De plus, dans le cas considéré, w^0 appartient à une et une seule composante de W^0 et est simple sur celle-ci.

Il reste à examiner le cas où w^0 est de la forme $\beta^0(s^0)$, le point s^0 étant un sommet de B^0. On se trouve alors dans le cas (**); compte tenu des hypothèses faites au n° 5, l'orbite de s^0 est composée de sommets deux à deux distincts. On peut prendre comme générateurs de l'idéal maximal $\mathfrak{m} = \mathfrak{m}(s^0, B)$ de l'anneau local $\mathfrak{o} = \mathfrak{o}(s^0, B)$ deux fonctions f_1 et f_2 sur B, définies sur K, et inversibles en tout sommet de B^0 distinct de s^0. Pour $\sigma \in \Gamma$, distinct de l'identité, les fonctions $f_i^{\sigma *}$ $(i=1, 2)$ sont inversibles en s^0. Donc si on pose $g_i = \prod_{\sigma \in \Gamma} f_i^{\sigma *}$, les fonctions f_1/g_1 et f_2/g_2 sont inversibles en s^0. Les fonctions g_1 et g_2 sont donc encore deux générateurs de $\mathfrak{m}$. Ces fonctions sont invariantes par Γ^*, donc sont respectivement de la forme $h_1 \circ \beta$ et $h_2 \circ \beta$, où h_1 et h_2 sont des fonctions sur W, définies sur k, et appartenant à l'idéal $\mathfrak{m}(w^0, W)$. On a donc $\mathfrak{m}(s^0, B) = \mathfrak{m}(w^0, W)\, \mathfrak{o}(s^0, B)$. Compte tenu de la $\mathfrak{p}$-normalité de W, ceci entraîne que s^0 est $\mathfrak{p}$-simple sur W. On a donc bien montré que W est $\mathfrak{p}$-simple.

Remarque. Le raisonnement ci-dessus montre en outre, dans le cas (**), que β, regardé comme morphisme de schémas sur R, est un *revêtement étale* au sens de [*4*]. On voit sans peine que la même propriété a lieu dans le cas (*) lorsque l'entier r est premier à la caractéristique p de k. J'ignore s'il en est encore ainsi dans le cas général.

Pour prouver la $\mathfrak{p}$-minimalité de W, il suffit d'appliquer (sous une forme légèrement modifiée) le critère utilisant la différentielle invariante donné dans [*6*], III, 2, prop. 1.

Les assertions du th. 1′ concernant le nombre des composantes de W^0 et leur connection se déduisent du caractère bijectif de la correspondance entre les orbites sur le support de B^0 et les points du support de W^0. Le fait que le coefficient de chacune de ces composantes est égal à l'indice séparable $r = r(V)$ se démontre par une étude élémentaire des propriétés des points de W algébriques sur k obtenus en relevant un point donné de supp W^0.

Enfin, compte tenu de la construction précédente, le th. 2, relatif au cas global, peut se déduire du th. 1, et de [*6*], II, 26, prop. 25, par un raisonnement analogue à celui utilisé dans [*6*], III, 16.

Bibliographie

[1] Châtelet, F.: Méthodes galoisiennes et courbes de genre 1. Annales de l'Université de Lyon, section A. **89**, 265–272 (1946).

[2] Châtelet, F.: Variation sur un thème de H. Poincaré. Annales de l'Ecole Normale Supérieure, **59**, 40–49 (1944).

[3] Dieudonné, J. et A. Grothendieck: Eléments de géométrie algébrique. Publ. Math. Inst. Htes. Et. Scientifiques, Paris, **4**, 1960.

[4] Grothendieck, H.: Séminaire de géométrie algébrique. Exposé I. Paris: I.H.E.S. 1960.

[5] Lang, S. and J. Tate: Principal homogeneous spaces over abelian varieties. Amer. J. Math., **78**, 659–684 (1956).

[6] Néron, A.: Modèles minimaux des variétés abéliennes sur les corps locaux et globaux. Publ. Math. Inst. Htes Et. Scient., Paris, **21** (1964).

[7] Ogg, A. P.: Cohomology of abelian varieties over function fields, Annals of Math., **76**, 2 185–212 (1962).

[8] Raynaud, M.: Caractéristique d'Euler-Poincaré d'un faisceau et cohomologie des variétés abéliennes. Séminaire Bourbaki, exposé 286 (1964–1965).

[9] Šafarevič, I. R.: Espaces principaux homogènes définis sur un corps de fonctions. Trud. Mat. Inst. V. A. Steklova, LXIV (en russe), ou Amer. Math. Soc. Transl., series 2, **37**, 85–114 (1964).

[10] Weil, A.: On algebraic groups and homogeneous spaces. Amer. J. Math., **77** 493–512 (1955).

[11] Weil, A.: Foundations of algebraic geometry. Amer. Math. Soc. Colloquium Publ., **29** 2nd edition, New York, 1946.

Passage au quotient par une relation d'équivalence plate

M. Raynaud

Cet exposé a pour but de justifier certains passages au quotient par des schémas en groupes *finis et plats*, non nécessairement étales, rencontrés par Tate dans sa théorie des groupes p-divisibles [*3*]. Nous montrons également, comment l'utilisation des faisceaux permet de préciser le problème des quotients en géométrie algébrique. L'auteur s'est directement inspiré de [*6*] Exp. IV et V.

1. Le point de vue classique

Soient k un corps, X une k-variété, sur laquelle opère un groupe algébrique G et cherchons à définir un quotient de X par G. Classiquement, on convient de prendre *l'espace annelé* $(X/G)_{\text{an}}$ quotient, dans la catégorie des espaces annelés, de X par la relation d'équivalence définie par l'action de G. Si $(X/G)_{\text{an}}$ est un préschéma, on obtient une pré-variété algébrique quotient. Ce point de vue présente les inconvénients suivants:

a) Si $(X/G)_{\text{an}}$ n'est pas un préschéma, l'espace annelé $(X/G)_{\text{an}}$ est pratiquement inutilisable en géométrie algébrique.

b) Même lorsque $(X/G)_{\text{an}}$ est un préschéma, on ne sait pas décrire pour autant le foncteur qu'il représente en terme de X et de G (c.à.d., on ne connait pas $\text{Hom}_k(Y,(X/G)_{\text{an}})$ pour tout k-préschéma Y).

Nous allons adopter ici un autre point de vue: au lieu de plonger les variétés, ou plus généralement les préschémas, dans la catégorie des espaces annelés, nous allons les plonger dans une catégorie plus étroitement liée à celle des préschémas et dans laquelle on dispose d'une théorie satisfaisante du passage au quotient. De telles catégories sont fournies par des *catégories de faisceaux* pour une topologie de Grothendieck convenable.

2. Choix de la topologie

Soient S un préschéma et Sch/S la catégorie des S-préschémas. En identifiant un préschéma X au foncteur qu'il représente:

$$T \rightsquigarrow \mathrm{Hom}_S(T, X) = X(T) \qquad T \in \mathrm{Sch}/S$$

on réalise Sch/S comme sous-catégorie pleine de la catégorie des préfaisceaux sur Sch/S. Soit $\mathcal{T}$ une topologie sur Sch/S. Un S-préschéma X est alors un faisceaux pour $\mathcal{T}$, si et seulement si, pour toute famille de morphismes de S-préschémas $T_i \to T (i \in I)$, qui est *couvrante* pour $\mathcal{T}$, le diagramme suivant est exact:

$$(*) \qquad X(T) \to \prod_i X(T_i) \rightrightarrows \prod_{i,j} X(T_i \times_T T_j).$$

La théorie de la déscente [5] nous dit que le diagramme (*) est exact si le morphisme $\coprod_{i \in I} T_i \to T$ est *fidèlement plat quasi-compact* (en abrégé: f.p.q.c.). Rappelons qu'un morphisme de préschémas $f: T' \to T$ est fidèlement plat si il est surjectif et si pour tout point t' de T', l'anneau local $0_{T',t'}$ est plat sur l'anneau $0_{T,f(t')}$; d'autre part, f est quasi-compact si l'image réciproque de tout ouvert affine de T est réunion d'un nombre fini d'ouverts affines de T' (condition toujours satisfaite si T' est noethérien). On définit la topologie f.p.q.c. sur Sch/S, comme étant la topologie engendrée par les familles surjectives d'immersions ouvertes et les morphismes f.p.q.c. (cf. [6] Exp. IV 6.3.). Désormais, nous supposons Sch/S munie de cette topologie; vu ce qui précède Sch/S est alors une sous-catégorie pleine de la catégorie des faisceaux pour la topologie f.p.q.c., de sorte que l'on a un diagramme commutatif:

$$\begin{array}{c} \mathrm{Sch}/S \hookrightarrow \text{Préfaisceaux} \\ \searrow \text{faisceaux} \nearrow \end{array}$$

3. Propriétés des quotients en théorie des faisceaux

Soient X un faisceau sur Sch/S, R un sous-faisceau de $X \times_S X$ qui est un *graphe d'équivalence* (c.à.d. pour tout S-préschéma T, $R(T)$ est le graphe d'une relation d'équivalence sur $X(T)$) et soient p_1 et p_2 les projections de R sur X. Alors on a les propriétés suivantes, valables en fait dans toute catégorie de faisceaux (cf. [6] Exp. IV):

i) Il existe un *faisceau conoyau*: $X \xrightarrow{p} X/R = Y$, de la double flèche $R \underset{p_1}{\overset{p_2}{\rightrightarrows}} X$.

Ce conoyau s'obtient de la manière habituelle en prenant le faisceau associé au préfaisceau conoyau (dont la définition est purement ensembliste).

ii) Le quotient $X \to Y$ est *effectif*, c.à.d.: le monomorphisme canonique

$$R \to X \times_Y X$$

est un *isomorphisme*.

iii) Le quotient $X \to Y$ est *universel*; celà signifie que pour tout faisceau Z et tout morphisme $Z \to Y$, le diagramme obtenu par changement de base:

$$R \times_Y Z \rightrightarrows X \times_Y Z \to Z$$

est encore *exact*.

4. Application aux préschémas quotients

Gardons les notations du numéro précédent mais supposons de plus que X soit un préschéma. *Le quotient de X par R* désignera toujours, dans la suite, le *faisceau quotient* $Y = X/R$, pour la topologie f.p.q.c. Si Y est représentable, nous dirons qu'il y a un *préschéma quotient* X/R (pour la topologie f.p.q.c.). Dans ce dernier cas, la traduction des propriétés i), ii) et iii) du N° 3 est immédiate et fournit des propriétés trés fortes (et trés utiles) du quotient X/R, propriétés qui ne sont pas vérifiées, en général, dans le cas classique rappelé au N° 1.

Proposition 1. Soient k un corps, X un k-préschéma, R une relation d'équivalence sur X. Alors, s'il existe un préschéma quotient X/R, pour la topologie f.p.q.c., X/R est aussi un quotient dans la catégorie des espaces annelés.

Celà tient au fait qu'un morphisme f.p.q.c. est submersif [*4*] chap. IV 2.3.12.

Remarque 1. Si Y est un préschéma quotient de X par R, il résulte du N° 3 ii) que R est isomorphe à $X \times_Y X$, de sorte que R est nécessairement *représentable* et que le monomorphisme $R \to X \times_S X$ est nécessairement une *immersion*. C'est là une restriction non négligeable sur la relation d'équivalence R. Ainsi, supposons qu'un S-schéma en groupes G opère librement sur un S-schéma X, de sorte que le morphisme u: $G \times_S X \to X \times_S X$

$$(g, x) \to (gx, x)$$

soit un monomorphisme et fasse de $G \times_S X$ le graphe d'une relation d'équivalence sur X. Mais u n'est pas nécessairement une immersion, et il peut arriver que le faisceau quotient X/G ne soit pas représentable, bien que l'espace annelè quotient $(X/G)_{an}$ soit un schéma (cf. [*8*] § 3 Example 04).

Toutefois, nous donnerons une réciproque partielle à la proposition 1 au N° suivant.

Proposition 2. Soit «P» une propriété de morphismes de préschémas qui soit stable par changement de base et de nature locale pour la topologie f.p.q.c. Alors, si $X \xrightarrow{p} Y$ est un préschéma quotient de X par une relation d'équivalence $R \underset{p_2}{\overset{p_1}{\rightrightarrows}} X$, *$p$ vérifie «P» si et seulement si p_1* (donc aussi p_2) vérifie «P».

En effet, d'après le N° 3 ii), le diagramme suivant est cartésien:

$$\begin{array}{ccc} X & \xleftarrow{p_2} & R \\ {\scriptstyle p}\downarrow & & \downarrow{\scriptstyle p_1} \\ Y & \xleftarrow{p} & X \end{array}$$

Donc, si p vérifie «P», p_1 vérifie «P». Par ailleurs, comme Y est un faisceau quotient de X pour la topologie f.p.q.c., il existe un diagramme commutatif:

$$\begin{array}{ccc} X & & \\ {\scriptstyle p}\downarrow & \nwarrow{\scriptstyle s} & \\ Y & \xleftarrow[g]{} & Y' \end{array}$$

où g est un morphisme f.p.q.c. Donc, si p_1 vérifie «P», le morphisme déduit de p par le changement de base f.p.q.c. $g: Y' \to Y$, vérifie «P», donc p vérifie «P».

L'intérêt de la proposition 2 provient du fait que les propriétés usuelles des morphismes, considérées en géométrie algébrique, satisfont aux conditions énoncées dans la proposition 2 (cf. [*4*] Chap. IV).

Exemples. a) Le morphisme p est *plat* si et seulement si p_1 est plat. Pratiquement, ce cas est le plus important.

b) Si p_1 est *lisse*, p est lisse et dans ce cas, Y représente aussi le faisceau quotient de X par R, lorsqu'on munit Sch/S de la topologie *étale*.

5. Théorèmes de représentabilité

Dans ce numéro, S est un préschéma localement noethérien, X un S-préschéma localement de type fini, R un *sous-préschéma* de $X x_S X$ qui est un graphe d'équivalence sur X, tel que la (première) projection $p_1: R \to X$ soit un *morphisme plat.* Alors:

Théorème 1. i) Il existe un plus grand ouvert U de X, tel que le quotient de U par la relation d'équivalence induite sur U par R soit représentable; U est dense dans X et si R est fermé dans $X x_S X$, U contient les points de X de codimension $\leqslant 1$.

ii) Soient S_0 un sous-préschéma fermé de S défini par un idéal nilpotent, $X_0 = X x_S S_0$, $R_0 = R x_S S_0$. Si X_0/R_0 est représentable, alors X/R est représentable.

iii) Si p_1 est *fini*, X/R est représentable si et seulement si toute classe d'équivalence suivant R, est contenue dans un ouvert affine.

Le quotient X/R est représentable dans chacun des cas suivants:

iv) X est quasi-projectif sur S et p_1 est *propre*.

v) X est *quasi-fini* sur S et R est fermé dans $X x_S X$

vi) Il existe un S-préschéma Z et un S-morphisme $u: X \to Z$, compatible avec R, tel que le morphisme canonique : $R \to X x_Z X$ soit surjectif.

Le résultat le plus important est l'assertion iii) du théorème 1; il est du à Grothendieck et sera démontré au numéro suivant. iv) résulte de iii) par une technique de quasi-sections ([6] Exp. V); ii) m'a été signalé par Grothendieck; i), v) et vi) figureront dans [9].

Corollaire 1. Soient k un corps, X un k-préschéma de type fini, R un *sous-préschéma* de $X x_k X$ qui est le graphe d'une relation d'équivalence sur X, telle que la première projection $p_1: R \to X$ soit un *morphisme plat*. Alors, les conditions suivantes sont équivalentes:

a) Le faisceau quotient X/R, pour la topologie f.p.q.c. est représentable.

b) L'espace annelé quotient $(X/R)_{\mathrm{an}}$ est un préschéma.

De plus les préschémas obtenus dans a) et b) sont isomorphes.

Exemples d'applications du théorème 1.

a) Soient G un S-schéma en groupes de type fini, H un sous-schéma en groupes fermé de G, plat sur S. Alors l'espace homogène quotient G/H est représentable dans chacun des cas suivants:

i) S est artinien [6] Exp. VI_{A}.

ii) G est lisse sur S et S régulier de dimension $\leqslant 1$ [9].

iii) G est quasi-projectif sur S et H est propre sur S (th. 1. iv)).

b) Grothendieck a représenté un ouvert convenable du foncteur de Picard relatif $Pic\, X/S$ d'un S-schéma projectif X, plat, à fibres géométriquement intègres, comme quotient d'un ouvert du schéma des diviseurs de X par une relation d'équivalence propre et plate [7].

c) Soient A un anneau de valuation discrète, K son corps de fractions, G une K-variété abélienne qui possède une bonne réduction sur A, de sorte que G se prolonge en un schéma abélien $\mathscr{G}$ sur $S=\operatorname{Spec} A$ et soit X un K-espace principal homogène sous G. Montrons que X se prolonge en un S-schéma *projectif* et *régulier* $\mathscr{X}$ (qui sera d'ailleurs un modèle minimal de X dans la terminologie de [*2*]). Soient K' une extension finie séparable de K qui trivialise X et S' le normalisé de S dans K. On a donc un K-morphisme:

$$u: G\times_K K' \xrightarrow{\sim} X\times_K K' \to X$$

Considérons le graphe d'équivalence de u : R, qui est un sous-schéma fermé de $G\times_K G\times_K K'\times_K K'$ et soit $\mathscr{R}$ l'adhérence de $\mathscr{R}$ dans $\mathscr{G}\times_S\mathscr{G}\times_S S'\times_S S'$. En utilisant l'action diagonale de $\mathscr{G}$ sur $\mathscr{G}\times_S\mathscr{G}$, on vérifie immédiatement que $\mathscr{R}$ est le graphe d'une relation d'équivalence sur $\mathscr{G}\times_S S'$, finie et plate. Soit $\mathscr{X}$ le schéma quotient. Comme le morphisme canonique $\mathscr{G}\times_S S'\to\mathscr{X}$ est f.p.q.c. et que $\mathscr{G}\times_S S'$ est régulier, $\mathscr{X}$ est régulier.

6. Démonstration du théorème 1. iii)

Commençons par traduire en terme de diagramme, le fait que R soit le graphe d'une relation d'équivalence sur X. Pour celà, considérons le sous-préschéma R de $X\times_S X$ comme X-préschéma grâce à la première projection p_1 et soit $R'=R\times_X R$ qui est canoniquement un sous-préschéma de $X\times_S X\times_S X$. On a donc un diagramme exact:

$$R' \underset{s_2}{\overset{s_1}{\rightrightarrows}} R \xrightarrow{p_1} X$$

$$(x_1, x_2, x_2') \begin{array}{l}\rightsquigarrow (x_1, x_2)\\ \rightsquigarrow (x_1, x_2')\end{array} \rightsquigarrow x_1 \quad (\forall\, T\in \mathrm{Sch}/S \quad \text{et} \quad x_1, x_2, x_2'\in X(T))$$

D'autre part, comme $R(T)$ est le graphe d'une relation d'équivalence sur $X(T)$, si (x_1, x_2) et $(x_1, x_2')\in R(T)$, on a aussi $(x_2, x_2')\in R(T)$, d'où une troisième flèche $s_3: R'\to R$

$$(x_1, x_2, x_2')\rightsquigarrow (x_2, x_2')$$

On vérifie alors, de manière purement ensembliste, que dans le diagramme suivant, le «carré supérieur» et le «carré inférieur» sont cartésiens:

$$\begin{array}{ccc} R' & \underset{s_2}{\overset{s_1}{\rightrightarrows}} & R \xrightarrow{p_1} X \\ {\scriptstyle s_3}\downarrow & & \downarrow{\scriptstyle p_2} \\ R & \underset{p_2}{\overset{p_1}{\rightrightarrows}} & X \end{array} \tag{1}$$

La démonstration du théorème 1. iii) se fait alors en quatre étapes:

a) Comme toute classe d'équivalence suivant R, est par hypothèse, contenue dans un ouvert affine de X et que p_1 est fini et plat, une réduction standard permet de nous ramener au cas où X est affine d'anneau A et où R est le spectre d'une A-algèbre B, libre de rang n sur A. Soit $B'=B\otimes_A B$. On déduit de

(1) le diagramme suivant, où les carrés sont cocartésiens:

$$\begin{array}{ccccc} B' & \overset{\sigma_1}{\underset{\sigma_2}{\leftleftarrows}} & B & \overset{\pi_1}{\leftarrow} & A \\ {\scriptstyle\sigma_3}\uparrow & & \uparrow{\scriptstyle\pi_2} & & \\ B & \overset{\pi_1}{\underset{\pi_2}{\leftleftarrows}} & A & & \end{array} \qquad (2)$$

Soit $A^0=(a\in A$ tels que $\pi_1(a)=\pi_2(a))$. On va bien sur montrer que le faisceau quotient X/R est représentable par $\operatorname{Spec}(A^0)$.

Dans toute la suite, nous considérons B comme A-algèbre *grâce au morphisme* π_2.

b) *A est entier sur A^0*. Soit $a\in A$ et $P(\pi_1(a))$ le polynôme caractéristique de l'élément $\pi_1(a)$ de la A-algèbre B. Si l'on fait l'un ou l'autre des changements de base $A\overset{\pi_1}{\underset{\pi_2}{\rightrightarrows}}B$, on trouve le polynôme caractéristique du même élément $\sigma_1\pi_1(a)=\sigma_2\pi_1(a)$ de la B-algèbre B' (considérée comme B-algèbre grâce à σ_3). On en déduit que les coéfficients de $P(\pi_1(a))$ sont dans A^0 et Hamilton-Cayley fournit une relation de dépendance intégrale de a sur A^0.

c) *Les éléments de A^0 «séparent» les classes d'équivalence de R dans X.* Celà résulte de b) et du théorème de Cohen-Seidenberg, par la considération de la *norme* d'un élément convenable.

d) Montrons que le morphisme $X\to\operatorname{Spec}(A^0)$ est fidèlement plat et que le morphisme canonique $R\to X\times_{\operatorname{Spec}(A^0)}X$ est un isomorphisme, ce qui achèvera la démonstration. Pour établir ce point, on peut supposer A^0 local, donc A semi-local (d'après c)). Comme R est un sous-schéma fermé de $X\times_S X$, le morphisme canonique: $A\otimes_{(A^0)}A\to B(a\otimes a'\rightsquigarrow\pi_1(a)\,\pi_2(a'))$ est surjectif. C'est dire que $\pi_1(A)$ (qui est un sous-A^0 module du A-module B), engendre le A-module B. Il résulte alors du lemme de Nakayama que $\pi_1(A)$ contient une base $\pi_1(c_1),\dots,\pi_1(c_n)$ du A-module B. Considérons alors le diagramme suivant:

$$\begin{array}{ccccc} B' & \overset{\sigma_1}{\underset{\sigma_2}{\leftleftarrows}} & B & \overset{\pi_1}{\leftarrow} & A \\ {\scriptstyle \varrho_2}\uparrow & & \uparrow{\scriptstyle \varrho_1} & & \uparrow{\scriptstyle \varrho_0} \\ B^n & \leftleftarrows & A^n & \leftarrow & (A^0)^n \end{array} \tag{3}$$

où les flèches sont définies de la manière suivante: i) la ligne inférieure est déduite de la façon évidente du diagramme exact: $B \overset{\pi_1}{\underset{\pi_2}{\leftleftarrows}} A \leftarrow A^0$.

ii) si e_i, $i=1, \ldots, n$, désigne la base canonique de $(A^0)^n$, on a posé:

$$\varrho_0\left(\sum e_j a_j^0\right) = \sum c_j a_j^0 \qquad (a_j^0 \in A^0)$$

$$\varrho_1\left(\sum e_j a_j\right) = \sum \pi_1(c_j)\, \pi_2(a_j) \qquad (a_j \in A)$$

$$\varrho_2\left(\sum e_j b_j\right) = \sum \left\{\begin{matrix} \sigma_1 \pi_1(c_j) \\ \| \\ \sigma_2 \pi_1(c_j) \end{matrix}\right\} \sigma_3(b_j) \qquad (b_j \in B)$$

Il est immédiat que (3) est commutatif et que ses deux lignes horizontales sont exactes. Par ailleurs, vu le choix des éléments c_i, ϱ_1 est un isomorphisme. Comme les carrés de (2) sont cocartésiens, on en déduit que ϱ_2 est aussi un isomorphisme, mais alors ϱ_0 est un isomorphisme c.q.f.d.

Remarque 2. Nous nous sommes volontairement limités, dans cet exposé, au cas d'une relation d'équivalence sur un préschéma, qui est donnée par son *graphe*. Pour étudier des situations plus générales, il est commode d'introduire le formalisme des groupoïdes; nous renvoyons le lecteur à [*6*] Exp. V.

Bibliographie

[*1*] Verdier, J. L.: A Duality Theorem in the Etale Cohomology of Schemes. Ce Volume p. 184–198.

[*2*] Néron, A.: Modèles minimaux des espaces homogènes. Ce Volume p. 66–77.

[*3*] Tate, J.: «*p*-divisible groups». Ce Volume p. 158–183.

[*4*] Grothendieck, A. et J. Dieudonné: Eléments de géométrie algébrique. I.H.E.S.

[*5*] Grothendieck, A.: Séminaire de géométrie algébrique 60/61. I.H.E.S.

[*6*] Demazure, M. et A. Grothendieck: Séminaire de géométrie algébrique 1963. Schémas en groupes. I.H.E.S.

[*7*] Grothendieck, A.: Schémas de Picard. Sem. Bourbaki 1961/62 No. 232.

[*8*] Mumford, D.: Géométric invariant theory. Berlin, Heidelberg, New York: Springer 1965.

[*9*] Raynaud, M.: Schémas en groupes sur un anneau de valuation discrète (à paraître).

Algebraische Aspekte in der nichtarchimedischen Analysis

R. Remmert

Seit Hensels bahnbrechenden zahlentheoretischen Untersuchungen über p-adische Zahlen zu Beginn dieses Jahrhunderts [*13*], [*14*] gibt es eine nichtarchimedische Analysis. Kürschak und Ostrowski klärten die bewertungstheoretischen Grundlagen; systematische Beiträge zur allgemeinen Funktionentheorie über einem nicht archimedisch bewerteten Körper gab es aber zunächst kaum. Die erste größere Arbeit scheint die Dissertation von W. Schöbe [*30*] aus dem Jahre 1930 zu sein, deren Resultate indessen lange unbekannt blieben und später von anderen – zum Teil auf komplizierterem Wege – neu bewiesen wurden. Weiterführende Ergebnisse verdankt man u.a. M. Krasner [*17*] und M. Lazard [*19*]. Eine einfache Herleitung der klassischen Resultate der Funktionentheorie einer Veränderlichen gab kürzlich U. Güntzer [*10*].

Im Winter 1961/62 hielt J. Tate [*33*] an der Harvard-University ein Seminar über „Rigid analytic spaces", durch das eine neue Entwicklung in der nichtarchimedischen Funktionentheorie eingeleitet wurde. Analog wie in der algebraischen und komplex-analytischen Geometrie werden Modellräume (affinoide Räume in der Terminologie von [*6*]) eingeführt und zu global-analytischen Räumen „zusammengeklebt"; da die Grundkörper total unzusammenhängend sind, muß das Prinzip der analytischen Fortsetzung per definitionem erzwungen werden. Die lokale Theorie, d.h. das Studium der Modellräume – oder äquivalent dazu: das Studium der Modellringe – ist gegenwärtig in stürmischer Entwicklung und erst partiell abgeschlossen. Im folgenden wird über einige algebraische Aspekte dieser Theorie berichtet, u.a. über zwei Arbeiten von L. Gerritzen [*4*], [*5*], in denen neue Methoden zur Beherrschung der Funktionentheorie über Grundkörpern k, deren Wurzelkörper $\sqrt[p]{k}$, $p := $ Charakteristik von k, unendlich-dimensionale k-Vektorräume sind, entwickelt werden.

§ 1. Normen. α-Orthogonalbasen in Vektorräumen

I bezeichnet stets einen kommutativen *nicht-trivial, nichtarchimedisch bewerteten* Integritätsring (*mit Eins*) mit multiplikativ geschriebener Bewertung $|\ |$, die nicht notwendig vollständig ist. K bezeichnet den Quotientenkörper von I mit der fortgesetzten Bewertung. M bezeichnet stets einen I-Modul, V stets einen K-Vektorraum.

1. Normen. – Eine Abbildung $\|\ \|: M \to \mathbb{R}$ heißt eine *I-Modulnorm auf M*, kurz: *Norm auf M*, wenn die folgenden Bedingungen erfüllt sind:

$$\|x\| \geqslant 0, \quad \text{Gleichheit gilt nur für} \quad x = 0 .$$
$$\|a \cdot x\| = |a| \cdot \|x\| \quad \text{für alle} \quad a \in I,\ x \in M .$$
$$\|x + y\| \leqslant \max(\|x\|, \|y\|) \quad \text{für alle} \quad x, y \in M .$$

Dann folgt (wie in der Bewertungstheorie):

$$\|x + y\| = \max(\|x\|, \|y\|), \quad \text{wenn} \quad \|x\| \neq \|y\| .$$

Die Werte $\|x\| \neq 0$ liegen nicht notwendig in der *Wertehalbgruppe* $|I^*|$ von $I^* := I - \{0\}$. Daher kann man, selbst wenn $|I^*|$ eine Gruppe ist, Vektoren $x \neq 0$ i.a. nicht durch Skalarenmultiplikation auf die Länge 1 eichen.

Jede Norm auf M macht M zu einem *topologischen hausdorffschen I-Modul*. Ein I-Modul M mit Norm heißt *normiert*; jeder normierter I-Modul ist *torsionsfrei*. Ein topologischer I-Modul M heißt *normierbar*, wenn die Topologie auf M von einer Norm herrührt.

Jede I-lineare *beschränkte* Abbildung $\varphi: M_1 \to M_2$ zwischen normierten I-Moduln M_1, M_2 ist *stetig*. Die Umkehrung ist *nicht* immer richtig, sie gilt aber stets dann, wenn es zwei Elemente $a, b \in I$ gibt mit $0 < |a| < 1 < |b|$. Ist diese Bedingung erfüllt, so sind speziell zwei Normen $\|\ \|_1$ und $\|\ \|_2$ auf M *äquivalent* (d.h. induzieren auf M dieselbe Topologie) genau dann, wenn es reelle Zahlen $\varrho, \varrho' > 0$ gibt, so daß für alle $x \in M$ gilt:

$$\varrho \|x\|_1 \leqslant \|x\|_2 \leqslant \varrho' \|x\|_1 .$$

Jede I-Modulnorm auf M ist in eindeutiger Weise zu einer K-Vektorraumnorm auf

$$M \otimes_I K = \left\{ \frac{x}{a},\ x \in M,\ a \in I^* \right\}$$

fortsetzbar. Jede beschränkte Linearform $\varphi: M \to I$ ist eindeutig zu einer beschränkten Linearform $\varphi: M \otimes_I K \to K$ (mit gleicher Schranke) fortsetzbar; stetige, nicht beschränkte Linearformen haben unstetige Fortsetzungen.

Jeder endlich-dimensionale topologische hausdorffsche K-Vektorraum V ist normierbar, (trägt i.a. aber nicht Produkttopologie!). Dies ergibt sich unmittelbar durch Übergang zur Komplettierung aus folgendem bekannten Satz:

Jeder endlich-dimensionale topologische hausdorffsche Vektorraum V über einem vollständig bewerteten Körper ist normiert, je zwei Normen auf V sind äquivalent. V trägt Produkttopologie und ist vollständig.

2. α-Orthogonalität. – Ist $\| \ \|$ eine Norm auf V, so gilt $\|v\| \leqslant \sup(|c_i| \cdot \|e_i\|)$ für jede konvergente Reihe $v = \sum_1^\infty c_i e_i$, $c_i \in K$, $e_i \in V$. Wichtig für Anwendungen sind Normabschätzungen nach unten.

Es sei α reell, $0 < \alpha \leqslant 1$. Eine Folge $\{e_1, e_2, \ldots\}$ von Elementen $e_i \neq 0$ aus V heißt *α-orthogonal*, wenn für jede konvergente Reihe $v = \sum_1^\infty c_i e_i$, $c_i \in K$, gilt:

$$\|v\| \geqslant \alpha \cdot \sup(|c_i| \cdot \|e_i\|);$$

alsdann sind die Vektoren $e_1, e_2, \ldots$ notwendig linear unabhängig.

Eine α-orthogonale Folge $\{e_1, e_2, \ldots\} \subset V$ heißt *α-Orthogonalbasis* von V, wenn jeder Vektor $v \in V$ sich als konvergente Reihe $\sum_1^\infty c_i e_i$ darstellen läßt; die Koordinaten $c_i \in K$ sind dann eindeutig durch v bestimmt (zur Motivierung dieses Begriffes, speziell für $\alpha = 1$, vgl. [*20*]).

Ein normierter, höchstens abzählbar-dimensionaler Vektorraum V besitzt i.a. zu keinem α, $0 < \alpha \leqslant 1$, eine α-Orthogonalbasis. Eine notwendige Bedingung ist, daß jeder endlichdimensionale Untervektorraum von V abgeschlossen ist.

Diese Bedingung ist auch hinreichend, genauer gilt:

Satz 1 ([4], [*26*]): *Es sei V ein höchstens abzählbar-dimensionaler K-Vektorraum und $\| \ \|$ eine Norm auf V, so daß jeder endlich-dimensionale Untervektorraum von V abgeschlossen ist. Dann gibt es zu jedem α, $0 < \alpha < 1$, eine α-Orthogonalbasis $\{e_1, e_2, \ldots\}$ von V.*

Der Beweis verläuft in Analogie zum E. Schmidtschen Orthogonalisierungsverfahren; man geht von irgendeiner Basis $\{v_1, v_2, \ldots\}$ von V aus und konstruiert eine α-Orthogonalbasis $\{e_1, e_2, \ldots\}$ sukzessive durch

Gleichungen

$$e_1 := v_1$$
$$e_2 := b_{21}v_1 + v_2$$
$$\vdots$$
$$e_n := b_{n1}v_1 + \cdots + b_{n,n-1}v_{n-1} + v_n$$

Anmerkung: Eine 1-Orthogonalbasis existiert unter den im Satz 1 gemachten Voraussetzungen i.a. nicht, vgl. [*21*].

Die Bedingung, daß jeder endlich-dimensionale Unterraum von V abgeschlossen ist, ist äquivalent damit, daß auf jedem endlich-dimensionalen Untervektorraum von V die Produkttopologie induziert wird. Dies ist wieder äquivalent damit, daß jeder endlich-dimensionale Untervektorraum U von V *stetig linear separiert* ist, d.h. daß zu jedem $v \neq 0$ aus U eine stetige Linearform $\lambda: U \to K$ mit $\lambda(v) \neq 0$ existiert.

Wir notieren einige wichtige Folgerungen aus Satz 1.

Korollar 1 ([*4*], p. 185): *Sei V wie im Satz 1. Dann ist V stetig linear separiert.*

Ist K vollständig bewertet, so ist jeder endlich-dimensionale K-Vektorraum vollständig, was zur Konsequenz hat, daß jeder endlich-dimensionale Unterraum eines normierten K-Vektorraumes abgeschlossen ist. Man erhält alsdann, wenn man einen topologischen Vektorraum, der einen dichten höchstens abzählbar-dimensionalen Unterraum enthält *von abzählbarem Typ* nennt:

Korollar 2 ([*4*], [*26*]): *Jeder normierte Vektorraum von abzählbarem Typ über einem vollständig bewerteten Körper besitzt für jedes* α, $0<\alpha<1$, *eine* α*-Orthogonalbasis, und ist speziell stetig linear separiert.*

Anmerkung: Ist K maximal vollständig, so gibt es sogar 1-Orthogonalbasen. Gleiches gilt, wenn die Bewertung von K diskret ist und jede monoton fallende Folge $\|x_\nu\|$, $x_\nu \in V$, eine Nullfolge ist ([*20*], vgl. auch [*21*]). Die Aussage des Korollars 2 reicht aus, um wichtige Ergebnisse aus [*31*] auf beliebige vollständig bewertete Grundkörper zu verallgemeinern.

Ein normierter, vollständiger Vektorraum V heißt ein *Banachraum*. Ist K vollständig bewertet, so ist der Raum $c_0(K)$ der Nullfolgen

$$c_0(K) = \{x = (x_1, x_2, \ldots),\ x_\nu \in K,\ \lim x_\nu = 0\}$$

ein Banachraum, wenn man durch

$$\|x\| := \max_\nu \{|x_\nu|\}$$

eine Norm auf $c_0(K)$ erklärt.

Korollar 3: *Jeder nicht endlich-dimensionale K-Banachraum von abzählbarem Typ über einem vollständig bewerteten Körper ist als topologischer Vektorraum homöomorph zum Raum $c_0(K)$.*

Speziell ist also ein abzählbar-dimensionaler normierter Vektorraum über einem vollständig bewerteten Körper niemals vollständig (Beispiel: der algebraische Abschluß $\mathbb{Q}_p$ des p-adischen Zahlkörpers $\mathbb{Q}_p$ ist nicht vollständig, da er eine abzählbare Basis, bestehend aus über $\mathbb{Q}$ algebraischen Zahlen, besitzt).

§ 2. Normierte Algebren. Potenz-multiplikative Normen

I und K haben dieselbe Bedeutung wie im § 1. $\hat{K}$ bezeichnet die Komplettierung von K. $J \supset I$ bezeichnet stets eine kommutative I-Algebra, J ist ein I-Modul.

1. Algebranormen. – Eine Norm $\| \ \|$ auf J – genauer: eine I-Algebranorm auf J – ist eine I-Modulnorm auf J, für welche zusätzlich gilt:

$$\|1\| = 1\,,$$
$$\|x \cdot y\| \leqslant \|x\| \cdot \|y\| \quad \textit{für alle} \quad x, y \in J\,.$$

Dann ist $\| \ \|$ eine *Fortsetzung* der Bewertung von I, die Multiplikation in J ist stetig.[1] Die Fortsetzung einer Norm von J auf die K-Algebra $J \otimes_I K$ ist eine K-Algebranorm auf $J \otimes_I K$.

J zusammen mit einer Norm heißt eine *normierte* I-Algebra. Eine normierte I-Algebra ist insbesondere eine *torsionsfreie, topologische, hausdorffsche* I-Algebra. Eine topologische I-Algebra J heißt *normierbar*, wenn die Topologie auf J von einer Norm herrührt.

Jede endlich-dimensionale hausdorffsche K-Algebra ist normierbar.

Eine normierte vollständige $\hat{K}$-Algebra heißt eine *$\hat{K}$-Banach-algebra*. Die Komplettierung $\hat{L}$ jeder normierten K-Algebra ist eine $\hat{K}$-Banachalgebra; $\hat{L}$ kann nilpotente Elemente $\neq 0$ haben, selbst dann, wenn L ein endlich-algebraischer Oberkörper von K ist.

Beispiele: Ist J eine normierte I-Algebra und $X=(X_1, \ldots, X_n)$ ein n-

[1] Die Multiplikation in J ist bzgl. einer I-Modulnorm $\| \ \|$ stets dann stetig, wenn es eine relle Zahl $\varrho \geqslant 1$ gibt, so daß für alle $x, y \in J$ gilt: $\|xy\| \leqslant \varrho \, \|x\| \cdot \|y\|$. Jede solche Norm läßt sich aber (durch Übergang zur Homothetienorm) durch eine äquivalente Norm ersetzen, die eine I-Algebranorm auf J ist.

tupel von Unbestimmten, so bildet die Gesamtheit

$$J\langle X\rangle := \{h = \sum_0^{\infty} a_\nu X^\nu,\ a_\nu \in J,\ a_\nu \to 0\}$$

aller *strikt konvergenten Potenzreihen über J* eine normierte I-Algebra; durch

$$\|h\| := \max_\nu \|a_\nu\| \tag{1}$$

wird die Norm von J auf $J\langle X\rangle$ fortgesetzt (dabei steht ν für das Tupel $\nu_1, \ldots, \nu_n$ und X^ν für $X_1^{\nu_1} \ldots X_n^{\nu_n}$).

Speziell ist $I\langle X\rangle$ eine normierte I-Algebra, die durch (1) definierte Norm ist eine Bewertung auf $I\langle X\rangle$ (Gauß'sches Lemma).

Die Gesamtheit

$$J\langle X, X^{-1}\rangle := \{h = \sum_{-\infty}^{+\infty} a_\nu X^\nu,\ a_\nu \in J,\ a_\nu \to 0\}$$

aller *strikt konvergenten Laurentreihen über J* ist ebenfalls vermöge (1) eine normierte I-Algebra.

Ist K vollständig und L eine K-Banachalgebra, so sind auch $L\langle X\rangle$ und $L\langle X, X^{-1}\rangle$ Banachalgebren.

Der Verzweigungsindex einer I-Norm $\| \ \|$ auf J wird analog wie in der Bewertungstheorie definiert: Die Menge $\|J \otimes_I K - \{0\}\| = \|J^*\| \cdot |K^*|$ besteht aus positiven reellen Zahlen und zerfällt, da sie die Wertegruppe $|K^*|$ von K umfaßt, in Äquivalenzklassen modulo $|K^*|$; die Anzahl dieser Klassen ist der *Verzweigungsindex e* der Norm $\| \ \|$. Es gilt:

$$e \leqslant \dim_K J \otimes_I K.$$

Bemerkung: $J \otimes_I K$ ist i.a. echt im vollen Quotientenring $Q(J)$ von J enthalten. Es gilt aber $J \otimes_I K = Q(J)$ stets dann, wenn J ganz über I ist.

2. *Die Funktoren $J \rightsquigarrow \mathring{J}$ und $J \rightsquigarrow \check{J}$.* – Für jede normierte I-Algebra J ist die Menge

$$\mathring{J} := \{x \in J : \text{die Folge } \{x^\nu\}_{\nu \geqslant 1} \text{ ist beschränkt in } J\}$$

der *potenz-beschränkten* Elemente von J ein Unterring von J. Es gilt

$$\mathring{I} = \{a \in I : |a| \leqslant 1\} \subset \mathring{J},$$

i.a. ist die Menge der $x \in J$ mit $\|x\| \leqslant 1$ *echt* in $\mathring{J}$ enthalten.

Da jeder beschränkte I-Algebrahomomorphismus potenz-beschränkte Elemente auf ebensolche abbildet, so ist die Zuordnung $J \rightsquigarrow \mathring{J}$ ein (kovarianter) Funktor der Kategorie der normierten I-Algebren mit den *beschränkten* I-Algebrahomomorphismen als Morphismen in die Kategorie der $\mathring{I}$-Algebren.

Die Menge

$$\mathfrak{t}(\mathring{J}) := \{x \in J : \text{die Folge } \{x^\nu\}_{\nu \geq 1} \text{ konvergiert gegen } 0\}$$

der *topologisch-nilpotenten* Elemente von J ist ein Ideal in $\mathring{J}$, das sein eigenes Radikal ist, $\mathfrak{t}(\mathring{J})$ besteht aus allen $x \in J$, zu denen ein Exponent m existiert mit $\|x^m\| < 1$ (die Menge aller x mit $\|x\| < 1$ ist i.a. *echt* in $\mathfrak{t}(\mathring{J})$ enthalten).

$\mathfrak{t}(\mathring{J}) \cap I = \mathfrak{t}(\mathring{I}) = \{a \in I, |a| < 1\}$ ist ein Primideal in I. Der Restklassenring

$$\tilde{J} := \mathring{J}/\mathfrak{t}(\mathring{J})$$

ist reduziert, die Injektion $\mathring{I} \hookrightarrow \mathring{J}$ induziert einen kanonischen Monomorphismus des *Integritätsringes* $\tilde{I}$ in $\tilde{J}$. Da stetige I-Algebrahomomorphismen auch topologisch-nilpotente Elemente auf ebensolche abbilden, ist die Zuordnung $J \rightsquigarrow \tilde{J}$ ein (kovarianter) Funktor der Kategorie der normierten I-Algebren mit den beschränkten I-Algebrahomomorphismen als Morphismen in die Kategorie der reduzierten $\tilde{I}$-Algebren.

Beispiele: Es gilt (Notationen wie in Abschnitt 1):

$\widetilde{I\langle X\rangle} = \tilde{I}[X] =$ Polynomalgebra in n Unbestimmten über $\tilde{I}$,

$\widetilde{I\langle X, X^{-1}\rangle} = \tilde{I}[X, X^{-1}] = \tilde{I}[X]_S$, wo $S = \{X_1^{\nu_1} \ldots X_n^{\nu_n}, \nu_1, \ldots, \nu_n \geq 0\}$.

Die $\tilde{I}$-Algebra $\tilde{J}$ ist torsionsfrei, denn aus $ax \in \mathfrak{t}(\mathring{J})$, $a \in I$, $|a| = 1$, $x \in J$, folgt stets $x \in \mathfrak{t}(\mathring{J})$, da $\|(ax)^\nu\| = \|x^\nu\|$ für alle $\nu \geq 1$ gilt. Der volle Quotientenring $Q(\tilde{J})$ von $\tilde{J}$ ist somit in natürlicher Weise eine Algebra über dem Quotientenkörper $Q(\tilde{I})$ von $\tilde{I}$[2]. Die Zahl

$$f := \dim_{Q(\tilde{I})} Q(\tilde{J})$$

heißt der *Restklassengrad* der normierten I-Algebra J. Äquivalente Normen ergeben denselben Restklassengrad, es gilt stets:

$$f \leq \dim_K J \otimes_I K.$$

In wichtigen Fällen ist $\tilde{J}$ noethersch. Z.B. gilt (vgl. hierzu [7], [*24*]):

Es sei J ein ganzer Oberring von I, so daß $Q(J)$ ein endlichdimensionaler Vektorraum über $Q(I)$ ist. $\tilde{I}$ sei noethersch und erweiterungsendlich.[3] *Dann ist $\tilde{J}$ ein noetherscher $\tilde{I}$-Modul.*

---

[2] Der volle Quotientenring eines kommutativen Ringes R wird stets mit $Q(R)$ bezeichnet.

[3] Zu diesem Begriff siehe § 4.

3. Potenz-multiplikative Normen. – Eine Norm auf J ist i.a. nicht äquivalent zu einer Bewertung. In wichtigen Fällen ist eine Norm $\| \ \|$ aber *potenzmultiplikativ*, d.h. es gilt

$$\|x^n\| = \|x\|^n \quad \text{für alle} \quad x \in J \quad \text{und alle} \quad n \geqslant 1 .$$

Solche Normen nennen wir abkürzend pm-*Normen.*

Die Fortsetzung einer pm-Norm von J auf $J \otimes_I K$ bleibt eine pm-Norm.

Für jede pm-Norm $\| \ \|$ auf J gilt

$$\mathring{J} = \{x \in J : \|x\| \leqslant 1\}, \ \mathfrak{t}(\mathring{J}) = \{x \in J : \|x\| < 1\} .$$

Ist der Verzweigungsindex e einer pm-Norm $\| \ \|$ endlich, so gilt

$$\|J\|^{e!} \subset |K| ,$$

denn von den $e+1$ Elementen $\|x^i\| = \|x\|^i \in \|J^*\| \cdot |K|$, $i = 0, 1, \ldots, e$, sind zwei modulo $|K^*|$ äquivalent.

Gibt es in $|I^*|$ Elemente <1 und >1, so sind äquivalente pm-Normen auf J stets gleich, denn aus

$$\varrho \, \|x^n\|_1 \leqslant \|x^n\|_2 \leqslant \varrho' \, \|x^n\|_1 , \qquad x \in J, \ \varrho, \varrho' \in \mathbb{R}$$

folgt durch Wurzelziehen, wenn es sich um pm-Normen handelt:

$$\sqrt[n]{\varrho} \, \|x\|_1 \leqslant \|x\|_2 \leqslant \sqrt[n]{\varrho'} \, \|x\|_1 \quad \text{für alle} \quad x \in J , \quad \text{d.h.} \quad \| \ \|_1 = \| \ \|_2 .$$

Sind $\| \ \|_1, \ldots, \| \ \|_s$ pm-Normen auf J, so wird durch

$$\|x\| := \max_{1 \leqslant j \leqslant s} \{\|x\|_j\} , \qquad x \in J ,$$

wieder eine pm-Norm gegeben. Wir nennen diese Norm eine *Maximumnorm auf J*, wenn $\| \ \|_1, \ldots, \| \ \|_s$ sämtlich Bewertungen auf J sind. Es gilt ([5]):

Satz 1: *Ist J nullteilerfrei und ganz über I und $Q(J)$ ein endlich-algebraischer Oberkörper von K, so ist jede* pm-*Norm* $\| \ \|$ *auf J eine Maximumnorm. Genauer: sind* $| \ |_i$, $i = 1, \ldots, s$, *die endlich vielen inäquivalenten Bewertungsfortsetzungen der Bewertung von K auf $Q(J)$, so gibt es eine Teilmenge* $\mathfrak{i} \neq \emptyset$ *von* $\mathfrak{s} := \{1, \ldots, s\}$, *so daß gilt:*

$$\|x\| = \|x\|_{\mathfrak{i}} := \max_{i \in \mathfrak{i}} \{|x|_i\} \quad \textit{für alle} \quad x \in J .$$

Je zwei Normen $\| \ \|_{\mathfrak{i}}$, $\| \ \|_{\mathfrak{i}'}$, mit $\mathfrak{i} \neq \mathfrak{i}'$ sind inäquivalent. Unter all diesen Normen induziert $\| \ \|_{\mathfrak{s}}$ die gröbste Topologie auf J; diese Norm läßt sich auch wie folgt beschreiben:

Ist $\| \ \|$ *eine* pm-*Norm auf* J*, deren Fortsetzung nach* $Q(J)$ *die Produkttopologie (als* K*-Vektorraum) induziert, so gilt:* $\| \ \| = \| \ \|_s$.

Die Norm $\| \ \|_s$ läßt sich einfach berechnen:

Ist $Y^b + a_1 \cdot Y^{b-1} + \cdots + a_b \in K[Y]$ *das Minimalpolynom von* $x \in J$ *über* K*, so gilt:*

$$\|x\|_s = \max_{1 \leq \beta \leq b} \sqrt[\beta]{|a_\beta|}.$$

4. Quasi-vollkommene Körper. – Jede pm-normierte K-Algebra L ist reduziert. Die Komplettierung $\bar{L}$ einer pm-normierten K-Algebra L ist pm-normiert, die Fortsetzung der Norm bleibt eine pm-Norm; speziell ist also $\bar{L}$ reduziert. Es gilt folgende Umkehrung ([5]):

Satz 2: *Ist* L *eine endlich-dimensionale topologische hausdorffsche* K*-Algebra und ist* $\bar{L}$ *reduziert, so ist* L pm-*normierbar.*

Bemerkung: Die Beweise der Sätze 1 und 2 werden simultan geführt. Es ist leicht zu sehen, daß $\bar{L}$ stets reduziert ist, wenn L zusätzlich separabel über K ist; daher ist in solchen Fällen L immer pm-normierbar. Gefährlich sind aber inseparable Erweiterungen von K.

Wir betrachten zunächst den Fall, daß L *quasi-galoisch* (= *normal*) über K ist. Der Fixkörper L^G der Galoisgruppe G von L über K ist rein inseparabel über K und L ist separabel über L^G. Die Bewertung von K setzt sich eindeutig zu einer Bewertung von L^G fort, jedoch ist L nicht notwendig eine topologische L^G-Algebra, d.h. die Injektion $L^G \subsetneq L$ des bewerteten Körpers L^G in L ist nicht notwendig stetig. Die *Stetigkeit* dieser Injektion ist also, da L separabel über L^G liegt, eine hinreichende (und auch notwendige) Bedingung dafür, daß L eine pm-normierbare K-Algebra ist, sie ist sicher dann erfüllt, wenn der K-Vektorraum L^G Produkttopologie trägt.

Wir nennen einen bewerteten Körper K der Charakteristik $p \neq 0$ *quasi-vollkommen* (bzgl. der Bewertung von K), wenn jeder endlich-dimensionale Untervektorraum des Wurzelkörpers $\sqrt[p]{K}$ von K, der vermöge der kanonischen Bewertungsfortsetzung ein normierter K-Vektorraum ist, Produkttopologie trägt. Dies trifft genau dann zu, wenn die Komplettierung $\bar{K}$ von K *separabel* über K ist, d.h. wenn $\bar{K} \otimes_K \sqrt[p]{K}$ reduziert ist. Jeder vollkommene Körper und jeder vollständig bewertete Körper ist quasi-vollkommen; Beispiele nicht quasi-vollkommener Körper erhält man, wenn man bemerkt, daß ein diskreter Bewertungsring B genau dann erweiterungsendlich ist, wenn sein Quotientenkörper $Q(B)$ quasi-vollkommen ist. Aus obigen Überlegungen folgt nun, da jeder Körper $L \supset K$ in einen Normalkörper einbettbar ist:

Satz 3: *Jeder endlich-dimensionale topologische hausdorffsche Oberkörper L eines quasi-vollkommenen Körpers K ist* pm-*normierbar.*

Hieraus folgt zusammen mit Satz 1:

Es sei I ein bewerteter Integritätsring mit quasi-vollkommenem Quotientenkörper K. Dann ist jede nullteilerfreie, normierte I-Algebra $J \supset I$, deren Quotientenkörper $Q(J)$ endlich-algebraisch über K ist, pm-*normierbar.*

§ 3. Strikt konvergente Potenzreihen. Rückertsche Theorie

k bezeichnet einen nicht trivial vollständig nicht archimedisch bewerteten Körper. $X=(X_1,\ldots,X_n)$ ist ein n-tupel von Unbestimmten, $n \geqslant 1$; im Abschnitt 2 schreiben wir durchweg Y für X_n. $E:=\{a \in k, |a| \leqslant 1\}$ ist der *Bewertungsring*, $\mathfrak{m}:=\{a \in k, |a|<1\}$ das *maximale Ideal* von E, $\kappa:=E/\mathfrak{m}$ der *Restklassenkörper* zu k.

1. Der Ring T_n. – Wir bezeichnen mit

$$T_n := k\langle X\rangle = k\langle X_1,\ldots,X_n\rangle = \{h=\sum_0^\infty a_\nu X^\nu,\ a_\nu \in k,\ a_\nu \to 0\}$$

den Ring T_n der strikt konvergenten Potenzreihen in $X_1,\ldots,X_n$ über k, vgl. § 2.1. Die Norm

$$\|h\| = \max\{|a_\nu|\}$$

ist eine Bewertung auf T_n und macht T_n zu einer k-Banachalgebra; die Polynomalgebra $k[X_1,\ldots,X_n]$ liegt dicht in T_n. Als k-Banachräume sind alle T_n isometrisch isomorph zum Raum $c_0(k)$ der Nullfolgen, als k-Algebren sind T_n und T_m nur dann isomorph, wenn $n=m$. Der Ring $\tilde{T}_n = \mathring{T}_n/\mathfrak{t}(\mathring{T}_n)$ identifiziert sich in natürlicher Weise mit dem Polynomring $\kappa[X_1,\ldots,X_n]$ über dem Restklassenkörper κ.

Jedes $h \in T_n$ definiert in natürlicher Weise eine k-wertige Funktion auf dem n-dimensionalen Einheitspolyzylinder „mit Rand“ $E^n \subset k^n$, die in E^n *analytisch*, d.h. um jeden Punkt von E^n in eine konvergente Potenzreihe entwickelbar ist. Man hat somit einen (wegen des Identitätssatzes für Potenzreihen) injektiven k-Algebrahomomorphismus $T_n \hookrightarrow \Gamma(E^n, \mathfrak{O})$ von T_n in die k-Algebra aller im E^n analytischen Funktionen. Konvergenz in der Banachtopologie bedeutet gleichmäßige Konvergenz auf E^n.

Die Ringe T_n und ihre Restklassenringe sind nach TATE [33] die *Modellringe* der nichtarchimedischen Funktionentheorie.[4] Das Ziel dieses Paragraphen ist, die Struktur von T_n, der später (§§ 5, 6) die Rolle des Ringes I der §§ 1, 2 übernimmt, zu beschreiben.

Wir definieren noch $T_0 := k$. Wir schreiben abkürzend T statt T_n und setzen

$$T' := k\langle X_1, \ldots, X_{n-1}\rangle \subset T, \quad \text{falls} \quad n \geqslant 1.$$

Jedes $h \in T$ schreibt sich eindeutig in der Form

$$h = \sum_0^\infty h_\nu X_n^\nu, \qquad h_\nu \in T'.$$

Es gilt $\|h\| = \max_{\nu \geq 0} \|h_\nu\|$.

2. Division mit Rest für T_n. – Die durch WEIERSTRASS und E. LASKER [18] eingeleitete Algebraisierung der lokalen komplexen Funktionentheorie wurde 1933 durch W. RÜCKERT in einer grundlegenden Arbeit [27] systematisch forgeführt. Als entscheidendes Hilfsmittel erwies sich die erstmals von H. SPÄTH [32] für den Ring der konvergenten Potenzreihen in n Unbestimmten über dem Körper der komplexen Zahlen bewiesene sog. Weierstrass'sche Formel, die für diesen Ring eine „Division mit Rest" liefert und einfache Induktionsbeweise für grundlegende Sätze der lokalen Funktionentheorie ermöglicht. Diese Rückertsche Theorie ist mutatis mutandis auch für die Ringe T_n durchführbar.

Zunächst sei an die Division mit Rest für Polynomringe erinnert. Ist R ein kommutativer Ring mit Einselement, z.B. $R = T' = k\langle X_1, \ldots, X_{n-1}\rangle$, und

$$g = g_0 + g_1 \cdot Y + \cdots + g_{s-1} \cdot Y^{s-1} + g_s \cdot Y^s \in R[Y]$$

ein Polynom in einer Unbestimmten Y über R, dessen höchster Koeffizient $g_s \in R$ eine Einheit in R ist, so gibt es zu jedem $h \in R[Y]$ eindeutig bestimmte Polynome q, $r \in R[Y]$, so daß gilt:

$$h = q \cdot g + r, \qquad \operatorname{grad} r < s.$$

[4] Neben den Ringen T_n kann man auch noch andere Modellringe studieren. Z.B. kann man den Ring W_n derjenigen konvergenten Potenzreihen $f = \sum_0^\infty a_{\nu_1 \ldots \nu_n} X_1^{\nu_1} \ldots X_n^{\nu_n}$ betrachten, zu denen es jeweils ein $r > 1$ gibt, so daß gilt: $\lim |a^{\nu_1 \ldots \nu_n}| \, r^{\nu_1 + \ldots + \nu_n} = 0$ (Ring der im „abgeschlossenen Einheitspolyzylinder strikt konvergenten Potenzreihenkeime"). Diese Ringe W_n wurden von G. WASHNITZER [35] betrachtet.

Eine entsprechende Aussage gilt auch für den Ring $T = T'\langle Y\rangle \supset T'[Y]$, wenn man für g „geeignete“ Potenzreihen wählt. Ein Element $g = \sum_0^\infty g_\nu Y^\nu \in T = k\langle X_1, \ldots, X_{n-1}, Y\rangle$ heiße *allgemein in* Y *von der Ordnung* $s \geqslant 0$, wenn g_s eine Einheit in T ist, derart, daß gilt:

$$\|g\| = \|g_s\|, \; \|g_\nu\| < \|g\| \quad \text{für alle} \quad \nu > s.$$

Dies bedeutet, wenn wir $\|g\| = 1$ annehmen und mit τ den Restklassenepimorphismus von $\mathring{T}$ auf die Polynomalgebra $\tilde{T} = \kappa[X_1, \ldots, X_{n-1}, Y]$ bezeichnen, gerade, daß $\tau(g) \in \kappa[X_1, \ldots, X_{n-1}][Y]$ ein Polynom s.ten Grades in Y ist, dessen höchster Koeffizient ein Element $\neq 0$ aus κ ist, und das somit als Divisorpolynom für die Division mit Rest im Polynomring $\kappa[X_1, \ldots, X_{n-1}][Y]$ geeignet ist. Im Beweise des folgenden Satzes wird diese Tatsache in etwas verfeinerter Form verwendet.

Satz 1 (Division mit Rest für T): *Es sei* $g \in T$ *allgemein in* Y *von der Ordnung* $s \geqslant 1$. *Dann gibt es zu jedem* $h \in T$ *eindeutig bestimmte Elemente* $q \in T$, $r \in T'[Y]$ *mit* grad $r < s$, *so daß gilt:*

$$h = q \cdot g + r \quad (\textit{Weierstrass'sche Formel}).$$

Es gelten die Abschätzungen

$$\|q\| \leqslant \|g\|^{-1} \cdot \|h\|, \quad \|r\| \leqslant \|h\|.$$

Der folgende *Beweis* ist eine leichte Variante des Beweises aus [*6*]. Wegen $\|T\| = |k|$ ist nur der Fall $\|g\| = \|h\| = 1$ zu diskutieren. Wir zeigen zunächst die Existenz von q und r. Sei $g = \sum_0^\infty g_\nu Y^\nu \in T'[Y]$. Dann gilt $|k| \ni \varrho := \max_{\nu > s} \|g_\nu\| < 1$. Wir wählen $c \neq 0$ in k, so daß $\varrho \leqslant |c| =: \varepsilon < 1$ (der Fall $\varrho = 0$ ist möglich). Man hat einen natürlichen Ringepimorphismus $\tau_\varepsilon : \mathring{T} \to \kappa_\varepsilon[X_1, \ldots, X_{n-1}][Y]$, wo $\kappa_\varepsilon = E/\mathfrak{m}_\varepsilon$ mit $\mathfrak{m}_\varepsilon := \{x \in E : |x| \leqslant \varepsilon\}$. Bezüglich Y ist $\tau_\varepsilon(g)$ ein Polynom vom Grade s, dessen höchster Koeffizient eine Einheit in $\kappa_\varepsilon[X_1, \ldots, X_{n-1}]$ ist, und das mithin als Divisorpolynom für die Division mit Rest in $\kappa_\varepsilon[X_1, \ldots, X_{n-1}][Y]$ geeignet ist.

Es genügt, in T Folgen $\{h_\nu\}$, $\{q_\nu\}$, $\{r_\nu\}$, $\nu \geqslant 0$, zu konstruieren, so daß

i) $h_0 := h$, $r_\nu \in T'[Y]$, grad $r_\nu < s$,
ii) $\|h_\nu\| \leqslant \varepsilon^\nu$, $\|q_\nu\| \leqslant \varepsilon^\nu$, $\|r_\nu\| \leqslant \varepsilon^\nu$,
iii) $h_{\nu+1} = h_\nu - q_\nu g - r_\nu$;

alsdann gilt nämlich

$$q := \sum_0^\infty q_\nu \in \mathring{T},\ r := \sum_0^\infty r_\nu \in T'[Y] \quad \text{mit} \quad \operatorname{grad} r < s,$$

und

$$h = h_0 = \sum_0^\infty (h_\nu - h_{\nu+1}) = \sum_0^\infty (q_\nu g + r_\nu) = qg + r.$$

Seien $q_{\nu-1}, r_{\nu-1}, h_\nu$ bereits konstruiert, $\nu \geqslant 0$. Dann gilt $c^{-\nu} h_\nu \in \mathring{T}$ wegen ii). Ist nun

$$\tau_\varepsilon(c^{-\nu}\quad {}_\nu) = \tilde{q}_\nu^* \tau_\varepsilon(g) + \tilde{r}_\nu^*, \quad \operatorname{grad} r_\nu^* < s,$$

die Division mit Rest von $\tau_\varepsilon(c^{-\nu} h_\nu)$ in $\kappa_\varepsilon[X_1, \dots, X_{n-1}][Y]$ bzgl. $\tau_\varepsilon(g)$ und sind $q_\nu^* \in \mathring{T}$, $r_\nu^* \in T'[Y]$ mit $\operatorname{grad} r_\nu^* < s$ Urbilder von $\tilde{q}_\nu^*$ und $\tilde{r}_\nu^*$ in $\mathring{T}$, so folgt wegen $\operatorname{Ker} \tau_\varepsilon = \{f \in T : \|f\| \leqslant \varepsilon\}$:

$$\|c^{-\nu} h_\nu - q_\nu^* g - r_\nu^*\| \leqslant \varepsilon.$$

Setzt man

$$q_\nu := c^\nu q_\nu^*, \quad r_\nu := c^\nu r_\nu, \quad h_{\nu+1} := h_\nu - q_\nu g - r_\nu = c^\nu (c^{-\nu} h_\nu - q_\nu^* g - r_\nu^*)$$

so gilt $\operatorname{grad} r_\nu < s$, $\|q_\nu\| \leqslant \varepsilon^\nu$, $\|r_\nu\| \leqslant \varepsilon^\nu$ und

$$\|h_{\nu+1}\| \leqslant \varepsilon^\nu \|c^{-\nu} h_\nu - q_\nu^* g - r_\nu^*\| \leqslant \varepsilon^{\nu+1},$$

d.h. i)–iii) sind für q_ν, r_ν und $h_{\nu+1}$ erfüllt. Damit ist der Existenzbeweis geführt; zugleich haben sich die behaupteten Abschätzungen ergeben, da $\|q\| \leqslant 1$ und analog $\|r\| \leqslant 1$.

Es bleibt zu zeigen, daß q und r eindeutig durch f bestimmt sind. Dazu ist nur zu beweisen, daß aus

$$q \cdot g + r = 0, \qquad q \in T, \quad r \in T'[Y], \quad \operatorname{grad} r < s$$

folgt: $q = r = 0$. Wäre $r \neq 0$, so dürfen wir $\|r\| = 1$ und also auch $\|q\| = 1$ annehmen. Es folgt

$$0 \neq \tau_\varepsilon(r) = -\tau_\varepsilon(q) \cdot \tau_\varepsilon(g) \in \kappa_\varepsilon[X_1, \dots, X_{n-1}][Y]$$

was aus Gradgründen unmöglich ist. – Satz 1 ist bewiesen.

3. Weierstrass'scher Vorbereitungssatz. Weierstrasspolynome. – Eine unmittelbare Anwendung der Division mit Rest liefert

Satz 2 (Weierstrass'scher Vorbereitungssatz): *Es sei $g \in T$ allgemein in Y von der Ordnung $s \geqslant 1$. Dann gibt es genau ein normiertes Polynom s.ten*

Grades $\omega \in T'[Y]$, *so daß gilt:*

$$g = e \cdot \omega \quad \text{mit einer Einheit} \quad e \in T.$$

Es bestehen die Gleichungen

$$\|e\| = \|g\|, \quad \|\omega\| = 1.$$

Falls $g \in T'[Y]$, *so gilt auch* $e \in T'[Y]$.

Beweis: Nach Satz 1 gilt

$$Y^s = q \cdot g + r \quad \text{mit} \quad r \in T'[Y], \quad \operatorname{grad} r < s, \quad \|r\| \leqslant \|Y^s\| = 1.$$

Folglich ist $\omega := Y^s - r \in T'[Y]$ ein Polynom s.ten Grades in Y mit $\|\omega\| = 1$, das allgemein in Y vom Grade s ist. Mithin gilt nach Satz 1 auch eine Gleichung

$$g = e \cdot \omega + \hat{r} \quad \text{mit} \quad \operatorname{grad} \hat{r} < s.$$

Zusammen mit $q \cdot g = \omega$ folgt $g = e(qg) + \hat{r}$, d.h.

$$0 = (eq - 1)\,g + \hat{r}.$$

Da dies die Division mit Rest von $f := 0$ bzgl. g ist, folgt $eq = 1$ aus der Eindeutigkeitsaussage von Satz 1. Also gilt $g = e\omega$ mit einer Einheit $e \in T$.

Ist $g = e' \cdot \omega'$ mit $\omega' = Y^s - r'$, grad $r' < s$ eine 2. Darstellung, so folgt $Y^s = e'^{-1} \cdot g + r'$ und also $r' = r$, d.h. $\omega' = \omega$ und $e' = e$ wieder nach Satz 1. Ist schließlich $g \in T'[Y]$, so gilt nach der Division mit Rest für Polynomringe eine Gleichung $g = p \cdot \omega + a$, wo $p \in T'[Y]$. Da auch $g = e \cdot \omega$ die Darstellung von g bzgl. ω gemäß Satz 1 ist, folgt $a = 0$ und $e = p$, w.z.b.w.

Bemerkung: Eine Weierstrass'sche Formel sowie ein Vorbereitungssatz gelten auch für den Ring $k\langle X, X^{-1}\rangle$ der strikt konvergenten Laurentreihen über k, wie GÜNTZER [*11*] zeigte; für k darf man dabei sogar beliebige vollständig filtrierte Ringe zulassen. Die Formel und der Vorbereitungssatz gelten ferner auch für die in Fußnote 4 beschriebenen Ringe W_n, vgl. [*12*].

Jedes normierte Polynom $\omega \in T'[Y]$ mit $\|\omega\| = 1$ heißt ein *Weierstrasspolynom in Y*. Diese Polynome ersetzen die „ausgezeichneten Pseudopolynome" der klassischen punktalen komplex-analytischen Geometrie. Dies beruht u.a. auf folgendem Lemma, welches Induktionsbeweise nach n ermöglicht.

Lemma: *Ist* $\omega \in T'[Y]$ *ein Weierstrasspolynom, so induziert die Injektion* $T'[Y] \subsetneq T$ *einen k-Algebraisomorphismus:*

$$T'[Y]/T'[Y] \cdot \omega \xrightarrow{\sim} T/T\omega.$$

Beweis: Die Injektion $T'[Y] \subsetneq T$ induziert jedenfalls einen k-Algebrahomomorphismus

$$\Omega: T'[Y]/T'[Y]\cdot\omega \to T/T\omega.$$

Nach Satz 2 gilt $T\omega \cap T'[Y] = T'[Y]\cdot\omega$, daher ist Ω injektiv. Da jedes $f \in T$ nach Satz 1 modulo $T\omega$ zu einem Polynom $r \in T'[Y]$ kongruent ist, ist Ω auch surjektiv, w.z.b.w.

Folgerung 1: *Der T'-Modul $T/T\omega$ ist kanonisch isomorph zum freien Modul*

$$bT', \quad wo \quad b := \operatorname{grad}\omega.$$

$1, Y, \ldots, Y^{b-1} \in T$ *geben zu einer T'-Modulbasis von $T/T\omega$ Anlaß.*

Folgerung 2: *Ein Weierstrasspolynom $\omega \in T'[Y]$ ist ein Primelement in $T'[Y]$ genau dann, wenn es ein Primelement in T ist.*

4. *Scherungssatz.* – Wir schreiben wieder X_n anstelle von Y. Nicht jedes Element $\neq 0$ aus T_n, $n > 1$, ist bzgl. einer der „Erzeugenden" $X_1, \ldots, X_n$ von T_n allgemein (z.B. das Produkt $X_1 \cdot X_2 \cdot \cdots \cdot X_n$ nicht).[5] Um die Sätze 1 und 2 sowie das Lemma mit seinen Folgerungen anwenden zu können, muß man daher (analog zum komplexen Fall) sicherstellen, daß sich jedes $h \neq 0$ aus T durch einen k-Algebraisomorphismus $T \to T$ allgemein machen läßt. Im komplexen Fall läßt sich dies bekanntlich stets durch lineare Automorphismen bewerkstelligen. Im nichtarchimedischen Fall kommt es hingegen entscheidend auf die Struktur des Restklassenkörpers κ von k an; nur wenn κ unendlich viele Elemente hat, kommt man mit linearen Automorphismen aus. Allgemein gilt (hinsichtlich einer präziseren Formulierung sowie des Beweises vgl. [*6*] p. 405–408).

Satz 3 (Scherungssatz): *Zu jedem $h \neq 0$ aus $T = k\langle X_1, \ldots, X_n \rangle$, $n \geqslant 2$, gibt es k-Algebraautomorphismen $\sigma: T \to T$, so daß $\sigma(h)$ bzgl. X_n allgemein ist; dabei läßt sich σ stets als „Scherung" wählen, d.h. es gilt*

$$\sigma(X_\nu) = X_\nu + X_n^{c_\nu}, \qquad \nu = 1, \ldots, n-1, \quad \sigma(X_n) = X_n$$

mit natürlichen Zahlen $c_1, \ldots, c_{n-1}$ als Exponenten.

5. *Folgerungen.* – Wir beweisen nun nach Rückertschem Vorbild

Satz 4: *T_n ist noethersch und faktoriell.*

Beweis: Beide Aussagen werden durch Induktion nach n verifiziert, der Induktionsbeginn $n = 0$ ist jeweils trivial. Sei $n > 0$, wir benutzen entscheidend, daß es nach dem Scherungssatz und Satz 2 zu jedem $h \neq 0$ aus T_n

---

[5] Für $n = 1$ ist jedes Element $\neq 0$ allgemein bzgl. X_1.

eine Scherung $\sigma: T_n \to T_n$ und ein Weierstrasspolynom $\omega \in T_{n-1}[X_n]$ gibt, so daß gilt: $T_n \cdot \sigma(h) = T_n \cdot \omega$.

a) Ein kommutativer Ring R ist noethersch, wenn es zu jedem $h \neq 0$ aus R einen Ringautomorphismus $\varrho: R \to R$ gibt, so daß der Restklassenring $R/R \cdot \varrho(h)$ noethersch ist. – Sei $R = T_n$. Auf Grund obiger Bemerkung ist nur zu zeigen, daß jeder Ring $T_n/T_n\omega$, wo $\omega \in T_{n-1}[X_n]$ ein Weierstrasspolynom ist, noethersch ist. Dies ist aber klar, da T_{n-1} nach Induktionsannahme noethersch und $T_n/T_n\omega$ nach Folgerung 1 des Lemmas ein endlich erzeugter T_{n-1}-Modul ist.

b) Ein kommutativer Integritätsring I ist faktoriell, wenn es zu jeder Nichteinheit $h \neq 0$ aus I einen Ringautomorphismus $\iota: I \to I$ gibt, so daß das Hauptideal $I \cdot \iota(h)$ das Produkt endlich vieler Primhauptideale ist. – Sei $I = T_n$. Es genügt wieder, Hauptideale $T_n\omega$, wo $\omega \in T_{n-1}[X_n]$ ein Weierstrasspolynom ist, zu betrachten. Nach Induktionsannahme ist T_{n-1} faktoriell, nach dem Gauß'schen Lemma bleiben Polynomringe über faktoriellen Ringen faktoriell. Mithin ist $T_{n-1}[X_n]$ faktoriell und es gilt eine Gleichung

$$\omega = \omega_1 \cdot \dots \cdot \omega_s$$

mit Primpolynomen $\omega_1, \dots, \omega_s \in T_{n-1}[X_n]$. Da ω ein Weierstrasspolynom ist, kann man auch alle ω_j als Weierstrasspolynome wählen. Nach Folgerung 2 des Lemmas sind dann alle Ideale $T_n\omega_j$ Primideale in T_n. Wegen $T_n\omega = (T_n\omega_1) \cdot \dots \cdot (T_n\omega_s)$ folgt die Behauptung, w.z.b.w.

Der Ring T_1 ist stets ein Hauptidealring, jedes Ideal wird von einem Polynom erzeugt. In T_n, $n \geqslant 2$, gibt es hingegen stets Ideale, die keine Polynome $\neq 0$ enthalten.

Bemerkung: 1) Die Induktionsschlüsse im Beweis von Satz 4 funktionieren auch für die Ringe $\mathring{T}_n = E\langle X_1, \dots, X_n \rangle$, da die Folgerungen 1 und 2 des Lemma sowie ein Scherungssatz auch für $\mathring{T}_n$ gelten. Daher gilt Satz 4 für alle Ringe $\mathring{T}_n$, $n \geqslant 0$, wenn er für den Bewertungsring $\mathring{T}_0 = E$ gilt. E ist aber genau dann noethersch (und faktoriell), wenn k diskret bewertet ist, d.h. wenn die Gruppe $|k^*|$ zyklisch ist. Somit gilt:

$\mathring{T}_n$, $n \geqslant 0$, ist noethersch und faktoriell genau dann, wenn k diskret bewertet ist.

2) Der Scherungssatz gilt auch für die Ringe W_n (vgl. Fußnote 4). Daraus läßt sich wegen $W_0 = k$ wieder folgern, daß auch alle Ringe W_n noethersch und faktoriell sind, vgl. [*12*].

H. Cartan hat in [*3*], p. 190ff. durch eine Verfeinerung des unter a) geführten Beweises gezeigt, daß in der komplexen Funktionentheorie alle Ideale im Stellenring der konvergenten Potenzreihen abgeschlossen bzgl.

einer natürlichen Topologie sind. Die entsprechenden Schlüsse sind auch für den Ring T_n durchführbar, sie liefern, wenn man für jede natürliche Zahl $j>1$ den freien T_n-Modul jT_n der geordneten j-tupel $(h_1,, \dots h_j)$ durch

$$\|(h_1, \dots, h_j)\| := \max_{1 \leqslant i \leqslant j} \{\|h_i\|\}$$

zu einem k-Banachraum macht:

Satz 5: *Jeder T_n-Untermodul $M \subset jT_n$ ist endlich erzeugbar. Zu jedem Erzeugendensystem $m_1, \dots, m_r$ von M gibt es eine reelle Schranke $R>0$, so daß zu jedem $m \in M$ Elemente $a_1, \dots, a_r \in T_n$ existieren, so daß gilt:*

$$m = \sum_1^r a_\varrho m_\varrho \quad \text{und} \quad \|a_\varrho\| \leqslant R\,\|m\|\,, \qquad \varrho = 1, \dots, r\,.$$

Zum Beweis vgl. [6], p. 411. Als Korollar erhält man ([6], p. 413):

Jeder T_n-Untermodul M von jT_n, $1 \leqslant j$, $n < \infty$, ist abgeschlossen.

Satz 5 und sein Korollar sind für ein tieferes Eindringen in die lokale nichtarchimedische Funktionentheorie nicht fein genug. Für viele Fragen sind die Ringe $\mathring{T}_n$ wichtiger als die Ringe T_n, und man benötigt, wenn $\mathring{T}_n$ nicht noethersch ist, Kriterien dafür, daß Untermoduln von $j\mathring{T}_n$ endlich erzeugbar und abgeschlossen sind. Ein erster Satz in dieser Richtung wurde in [7], § 2, hergeleitet:

Ein $\mathring{T}_n$-Untermodul $\mathring{M}$ von $j\mathring{T}_n$, $1 \leqslant j$, $n < \infty$, ist stets dann endlich erzeugbar (und abgeschlossen), wenn er saturiert ist, d.h. wenn gilt: $\mathring{M} \cdot T_n \cap j\mathring{T}_n = \mathring{M}$.

Zum Beweis wird im wesentlichen gezeigt, daß man im Satz 5 die Erzeugendensysteme so wählen kann, daß die Schranke R zu 1 wird; das entscheidende Hilfsmittel dazu bilden die in [7], § 1, eingeführten *quasi-noetherschen* B-Ringe. (Ein einfacher Beweis für diese Endlichkeitsaussage, der nur die Weierstrass'sche Formel benutzt, wurde inzwischen von L. GERRITZEN and U. GÜNTZER angegeben in: Über Restklassennormen auf affinoiden Algebren. Inv. Math. **3**, 71–74 (1967).)

Ein weiterer wichtiger Satz besagt, daß der Ring T_n erweiterungsendlich ist. Der Beweis, der im Falle $\operatorname{char} k = p \neq 0$ vor allem dann, wenn $\sqrt[p]{k}$ unendlich-dimensional über k ist, Schwierigkeiten macht, wird im nächsten Paragraphen besprochen.

§ 4. Erweiterungsendlichkeit und Quasivollkommenheit von T_n

I bezeichnet einen Integritätsring (ohne Bewertung), K seinen Quotientenkörper.

1. Noethersche erweiterungsendliche Ringe. – I heißt *erweiterungsendlich*, wenn in jedem endlich-algebraischen Oberkörper K' von K der ganze Abschluß I' von I ein endlicher I-Modul ist. Die Wichtigkeit dieses Begriffes wurde schon von E. NOETHER [*25*] erkannt; später haben F. K. SCHMIDT [*29*] und M. NAGATA [*23*] solche Ringe betrachtet[6] (vgl. auch [8]).

Ist K_a ein algebraischer Abschluß von K, so ist der ganze Abschluß I_a von I in K_a eine I-Algebra und bis auf Isomorphie eindeutig bestimmt; I_a heißt ein *universell-ganzer Abschluß von I.* Nun gilt:

Erstes Kriterium für Erweiterungsendlichkeit ([*4*]), p. 182): *Ein noetherscher Integritätsring I ist genau dann erweiterungsendlich, wenn jeder endlich-rangige I-Untermodul von I_a noethersch ist.*[7]

2. Normale Noethersche erweiterungsendliche Ringe. – Das 1. Kriterium für Erweiterungsendlichkeit kann wesentlich verbessert werden, wenn I zusätzlich normal (= ganz-abgeschlossen in K) ist. Dann sind nämlich, da nach R. DEDEKIND und E. NOETHER ([*25*], p. 36) in diesem Falle der ganze Abschluß von I in jeder endlich-algebraischen separablen Erweiterung K' von K ein noetherscher I-Modul ist, nur rein inseparable Oberkörper von K gefährlich, d.h. höchstens im Falle $\operatorname{char} I =: p \neq 0$ kann ein normaler noetherscher Ring I nicht erweiterungsendlich sein. Hier gibt es nun, wie F. K. SCHMIDT 1936 zeigte ([*29*], p. 444), sogar diskrete Bewertungsringe, die nicht erweiterungsendlich sind. Dies liegt daran, daß der I umfassende „Wurzelring“

$$\sqrt[p]{I} := \{x \in I_a : x^p \in I\}$$

nicht noethersche I-Untermoduln von endlichem Rang enthalten kann.

---

[6] Erweiterungsendliche Ringe werden in [*9*], p. 213 „anneaux japonais“ genannt. Dort wird auch folgendes von TATE, [*33*], p. 16, stammendes Kriterium für Erweiterungsendlichkeit bewiesen:

Ein normaler noetherscher Integritätsring I ist erweiterungsendlich, wenn es ein $x \neq 0$ in I gibt mit folgenden Eigenschaften:

1. *Ix ist ein Primideal, I ist hausdorffsch und vollständig für die x-adische Topologie.*
2. *Der Integritätsring I/Ix ist erweiterungsendlich.*

[7] Unter dem *Rang* eines I-Moduls M, in Zeichen $\operatorname{rg} M$, verstehen wir die Maximalzahl linear unabhängiger Elemente in M. Ist M endlich erzeugt, so gilt stets $\operatorname{rg} M < \infty$, die Umkehrung ist i.a. falsch.

Für jedes epimorphe Bild M' von M gilt: $\operatorname{rg} M' \leqslant \operatorname{rg} M$, woraus speziell folgt: Zu jedem Untermodul N einer direkten Summe $M_1 \oplus M_2$ gibt es einen Untermodul N_ν von M_ν, $\nu = 1,2$, sodaß gilt:

$$N \subset N_1 \oplus N_2\,, \ \operatorname{rg} N_\nu \leqslant \operatorname{rg} N,\ \nu = 1,2.$$

Es gilt nämlich

Zweites Kriterium für Erweiterungsendlichkeit ([*4*], p. 183): *Ein normaler noetherscher Integritätsring I mit* char $I=p\neq 0$ *ist genau dann erweiterungsendlich, wenn jeder endlichrangige I-Untermodul des Wurzelringes* $\sqrt[p]{I}$ *noethersch ist.*

Die Bedingung ist z.B. stets erfüllt, wenn $\sqrt[p]{I}$ selbst ein endlicher I-Modul ist (vgl. [*1*] sowie [*34*], p. 123).

Für Anwendungen nützlich ist folgendes

Lemma: *Es sei M ein Modul über einem noetherschen Integritätsring I, derart, daß jeder endlich-rangige Untermodul von M linear separiert ist.*[8] *Dann ist jeder endlichrangige Untermodul von M noethersch.*

Beweis (vgl. [*4*]): Sei $N\neq 0$ ein Untermodul von M vom Range t. Es gilt $t\geqslant 1$, da M torsionsfrei ist. Nach Voraussetzung gibt es ein $\lambda\in\mathrm{Hom}(N, I)$ mit $\mathrm{Ker}\,\lambda\neq N$. Es folgt $\mathrm{rg}(\mathrm{Ker}\,\lambda)<t$ (vgl. z.B. [*4*], § 1, Satz 2,ii)); nach Induktionsannahme ist also Ker λ noethersch (im Induktionsbeginn $t=0$ gilt $\mathrm{Ker}\,\lambda=0$). Da auch Bild $\lambda\subset I$ noethersch ist, muß N selbst noethersch sein, w.z.b.w.

3. Die Sätze von F. K. Schmidt und Nagata. – Als Anwendung zeigen wir den bekannten

Satz 1: *Für jeden Körper k ist der Ring* $k[X_1,\ldots, X_n]$ *der Polynome und der Ring* $k[\![X_1,\ldots, X_n]\!]$ *der formalen Potenzreihen in endlich vielen Unbestimmten erweiterungsendlich.*

Bemerkung: Für Polynomringe stammt dieser Satz von F. K. Schmidt ([*29*], p. 444); für formale Potenzreihen wurde er von M. Nagata [*23*] und Y. Mori [*22*] bewiesen.

Beweis des Satzes: Wir führen die Schlüsse für $I:=k[\![X_1,\ldots, X_n]\!]$ durch. Nach bekannten Sätzen ist I noethersch und faktoriell, also auch normal. Daher genügt es nach dem 2. Kriterium und dem Lemma zu zeigen, daß der I-Modul $\sqrt[p]{I}=\sqrt[p]{k}[\![\sqrt[p]{X_1},\ldots,\sqrt[p]{X_n}]\!]$ linear separiert ist. $\sqrt[p]{I}$ ist isomorph zu einer endlichen direkten Summe des I-Moduls

$$I^*:=\sqrt[p]{k}[\![X_1,\ldots,X_n]\!],$$

(die Monome $\sqrt[p]{X_1^{\nu_1}}\cdot\ldots\cdot\sqrt[p]{X_n^{\nu_n}}$, $0\leqslant\nu_i\leqslant p-1$, bilden eine I^*-Modulbasis),

[8] Ein I-Modul M heißt *linear separiert*, wenn es zu jedem $x\neq 0$ aus M ein $\lambda\in\mathrm{Hom}(M, I)$ mit $\lambda(x)\neq 0$ gibt. Solche Moduln sind stets torsionsfrei. Jeder Untermodul eines linear separierten Moduls ist linear separiert.

daher ist zu zeigen, daß I^* linear separiert ist. Sei also

$$0 \neq h = \sum_0^\infty a_\nu X^\nu \in I^*, \quad a_\nu \in \sqrt[p]{k}, \quad \text{etwa} \quad a_i \neq 0.$$

Da jedenfalls $\sqrt[p]{k}$ als k-Vektorraum linear separiert ist, gibt es eine k-lineare Abbildung $\lambda': \sqrt[p]{k} \to k$ mit $\lambda'(a_i) \neq 0$. Durch

$$\lambda(f) := \sum_0^\infty \lambda'(b_\nu) X^\nu \in I, \quad \text{falls} \quad f = \sum_0^\infty b_\nu X^\nu \in I^*,$$

wird λ' zu einer I-linearen Abbildung $\lambda: I^* \to I$ fortgesetzt. Da $\lambda(h) \neq 0$ nach Konstruktion von λ', so ist I^* linear separiert, w.z.b.w.

Vorstehender Beweis funktioniert wörtlich für Polynomringe, man schreibe einfach überall eckige Klammern.

4. Erweiterungsendlichkeit von T_n. – In diesem Abschnitt ist k ein vollständig bewerteter Körper, $\operatorname{char} k = p \neq 0$. Der soeben gegebene Beweis kann unter Heranziehung der Sätze aus § 1 zu einem Beweis für die Erweiterungsendlichkeit von $T = T_n = k\langle X_1, \ldots, X_n\rangle$, $n \geq 1$, ausgebaut werden. Entscheidend ist folgender

Hilfssatz: *Ist $V \subset \sqrt[p]{k}$ ein k-Vektorraum von abzählbarem Typ, so ist der normierte T-Modul*

$$M_V := \{h = \sum_0^\infty v_\nu X^\nu, \; v_\nu \in V, \; v_\nu \to 0\}$$

stetig linear separiert.

Beweis: Jede stetige Linearform $\lambda': V \to k$ ist beschränkt und induziert vermöge

$$\sum_0^\infty v_\nu X^\nu \to \sum_0^\infty \lambda'(v_\nu) X^\nu$$

eine T-lineare Abbildung $\lambda: M_V \to T$, da mit $v_\nu \to 0$ wegen $|\lambda'(v_\nu)| \leq C|v_\nu|$ auch $\lambda'(v_\nu) \in k$ gegen 0 konvergiert. Es gilt $\|\lambda(h)\| \leq C\|h\|$, d.h. λ ist stetig.

Da V nach Korollar 2 zu Satz 1.2 ein stetig linear separierter k-Vektorraum ist, folgt (wie im Beweis von Satz 1), daß M_V ein stetig linear separierter T-Modul ist, w.z.b.w.

Korollar: *Jeder endlich-rangige T-Untermodul von $\sqrt[p]{T} = \sqrt[p]{k}\langle \sqrt[p]{X_1}, \ldots, \sqrt[p]{X_n}\rangle$ ist stetig linear separiert.*

Beweis: $\sqrt[p]{T}$ ist isomorph zu einer endlichen direkten Summe des T-Moduls

$$T^* := \sqrt[p]{k}\langle X_1, \ldots, X_n\rangle;$$

daher genügt es zu zeigen, daß jeder endlich-rangige T-Untermodul $N \neq (0)$ von T^* linear separiert ist. Seien

$$f_i = \sum_0^\infty a_{i\nu} X^\nu, \qquad i = 1, \ldots, t$$

maximal viele T-linear unabhängige Elemente aus N. Die abzählbare Koeffizientenmente $\{a_{i\nu}\}$, $i=1,\ldots,t$, $\nu \geqslant 0$, erzeugt einen k-Vektorraum $V \subset \sqrt[p]{k}$ von abzählbarem Typ ($\sqrt[p]{k}$ selbst ist nicht notwendig von abzählbarem Typ über k), nach dem Hilfssatz ist der T-Modul $M_V = \{h = \sum_0^\infty v_\nu X^\nu, v_\nu \in V, v_\nu \to 0\}$ linear separiert. Es gibt aber $N \subset M_V$. Ist nämlich $f \in N$ beliebig, so gibt es nach Wahl der $f_1, \ldots, f_t$ Elemente $g, g_1, \ldots, g_t \in T$ mit $g \neq 0$, so daß gilt

$$gf = \sum_1^t g_i f_i \in M_V, \quad \text{und also auch} \quad f \in M_V.$$

Mit M_V ist also auch $N \subset M_V$ ein linear separierter T-Modul, w.z.b.w.

Da T noethersch und faktoriell ist (Satz 3.4), folgt jetzt unmittelbar aus dem Lemma und dem 2. Kriterium für Erweiterungsendlichkeit:

Satz 2 ([*4*], p. 187): *Für jeden vollständig bewerteten Körper k ist der Ring $T = k\langle X_1, \ldots, X_n\rangle$ der strikt konvergenten Potenzreihen in n Unbestimmten über k erweiterungsendlich.*

Der obige Beweis funktioniert auch für den Ring $\mathring{T} = E\langle X_1, \ldots, X_n\rangle$, wenn k diskret bewertet ist (dann ist nämlich $\mathring{T}$ noethersch und faktoriell, und die Schranken aller verwendeten Linearformen darf man offensichtlich als $\leqslant 1$ annehmen). Somit folgt noch:

Ist k diskret bewertet, so ist $\mathring{T} = E\langle X_1, \ldots, X_n\rangle$ erweiterungsendlich.[9]

Der dargestellte Beweis liefert mutatis mutandis auch, daß der Ring *aller* konvergenten Potenzreihen in n Unbestimmten über k erweiterungsendlich ist. Dabei kann man die Voraussetzung, daß k vollständig ist, abschwächen und zeigen:

Satz 3 ([*4*], p. 189): *Der Ring K_n der konvergenten Potenzreihen in n Unbestimmten, $n \geqslant 1$, über einem bewerteten Körper k ist genau dann erweiterungsendlich, wenn k quasi-vollkommen ist.*

5. *Quasivollkommenheit von $Q(T)$.* – Der Quotientenkörper $Q(T)$, $n \geqslant 1$, ist bzgl. der fortgesetzten Bewertung *nicht* vollständig. Um im nächsten

---

[9] Die Methoden von Gerritzen wurden inzwischen von R. Kiehl [*16*] so ausgebaut, daß sich bei vollständig bewerteten Grundkörpern k sogar die Exzellenz (im Sinne von Grothendieck) der Ringe T_n und K_n ergibt.

Paragraphen die Resultate des § 2, insbesondere Satz 2.3 und seine Folgerung, anwenden zu können, bemerken wir hier noch:

Satz 4 ([5]): *$Q(T)$ ist quasi-vollkommen.*

Beweis: Es genügt zu zeigen, daß jeder endlich-dimensionale $Q(T)$-Untervektorraum von $\sqrt[p]{Q(T)}=Q(\sqrt[p]{T})$ stetig linear separiert ist. Das ist sicher dann der Fall, wenn jeder endlich-erzeugte T-Untermodul von $\sqrt[p]{T}$ stetig linear separiert ist. Das folgt aber unmittelbar aus dem Korollar des Hilfssatzes, w.z.b.w.

§ 5. k-affinoide Algebren. Spektralnorm

k bezeichnet stets einen nichttrivial, nichtarchimedisch und vollständig bewerteten Grundkörper.

1. k-affinoide Algebren. – Eine k-Banachalgebra A heißt eine *k-affinoide Algebra*, wenn es endlich viele potenzbeschränkte Elemente $f_1,\dots,f_n \in \mathring{A}$ gibt, so daß sich jedes $f \in A$ als eine strikt konvergente Potenzreihe in den $f_1,\dots,f_n$ schreibt[10]; die Elemente $f_1,\dots,f_n \in \mathring{A}$ heißen alsdann ein *topologisches Erzeugendensystem* von A, und wir schreiben $A=k\langle f_1,\dots,f_n\rangle$. Die einfachste k-affinoide Algebra ist die „freie" Algebra $T_n=k\langle X_1,\dots, X_n\rangle$ in n Unbestimmten über k.

Ist $\{f_1,\dots,f_n\}$ ein topologisches Erzeugendensystem von A, so gibt es einen kanonischen stetigen Epimorphismus $T_n \to A$, der X_ν auf f_ν abbildet, $\nu=1,\dots,n$. Jede k-affinoide Algebra A ist mithin topologisch isomorph zu einer Restklassenalgebra $T_n/\mathfrak{a}$ und umgekehrt; versieht man $T_n/\mathfrak{a}$ mit der Restklassennorm, so ist $T_n \to T_n/\mathfrak{a}$ eine Kontraktion. Jede k-affinoide Algebra ist noethersch; jeder k-Algebrahomomorphismus $\varphi: A \to B$ zwischen k-affinoiden Algebren A, B ist stetig ([33], theorem 4.6; vgl. auch [6], p. 448 sowie Abschnitt 5 dieses §).

Ist A eine k-affinoide Algebra, so trägt jeder endlich erzeugte A-Modul M genau eine Banachtopologie, sodaß M ein topologischer A-Modul ist. Jeder A-Untermodul von M ist abgeschlossen in M; dies folgt leicht aus der entsprechenden Aussage für Untermoduln von jT_n, p. 102. Jeder A-

[10] Diese Definition ist „konstruktiv". Erwünscht wäre eine einfache „axiomatische" Charakterisierung der k-affinoiden Algebren. Der naheliegende Versuch, sie etwa als noethersche k-Banachalgebren zu definieren, in denen ein endlich erzeugter Polynomring dicht liegt und deren zugehörigen κ-Algebren κ-affin sind, führt nicht zum Ziele (vgl. [12]), wie der Ring der im Kreis vom Radius $\varrho > 0$, $\varrho \notin |\bar{k}|$, strikt konvergenten Potenzreihen zeigt.

Modulhomomorphismus $\lambda: M \to N$ zwischen endlich erzeugten A-Moduln M, N ist stetig; λ ist offen, falls $\lambda(M) = N$.

Beispiele k-affinoider Algebren: 1) Der Ring L_n der strikt konvergenten Laurentreihen in n Unbestimmten ist k-affinoid. Es gilt

$$L_n = k\langle X_1, Y_1, \dots, X_n, Y_n\rangle/\mathfrak{a}, \quad \text{wo} \quad \mathfrak{a} := (X_1 Y_1 - 1, \dots, X_n Y_n - 1).$$

Die natürliche Norm auf L_n ist gleich der Restklassennorm.

2) Ist A eine k-affinoide Algebra und sind $Y_1, \dots, Y_m$ Unbestimmte, so ist auch $A\langle Y_1, \dots, Y_m\rangle$ k-affinoid (vgl. § 2.1).

3) Jede Restklassenalgebra $A/\mathfrak{a}$ einer k-affinoiden Algebra A nach einem Ideal $\mathfrak{a} \neq A$ ist k-affinoid.

4) Jede k-Oberalgebra B einer k-affinoiden Algebra A, die ein endlicher A-Modul ist, ist eine k-affinoide Algebra.

2. Normalisierungslemma. – Grundlegend für die Theorie der k-affinoiden Algebren ist die Umkehrung der Aussage des Beispiels 4). Nennt man allgemein einen Homomorphismus $\varphi: A \to B$ zwischen kommutativen Ringen A, B *endlich*, wenn B bzgl. φ ein endlich-erzeugter A-Modul ist, so gilt:

Satz 1 (Normalisierungslemma): *Zu jeder k-affinoiden Algebra B gibt es eine natürliche Zahl $d \geqslant 0$ und einen endlichen k-Algebramonomorphismus $T_d \subsetneq B$; d ist die Krullsche Dimension von B.*

Beweis: Sei etwa $B = k\langle f_1, \dots, f_n\rangle$. Wir führen Induktion nach n, der Induktionsbeginn $n = 0$ ist trivial. Sei $n \geqslant 1$. Es gibt einen Epimorphismus $\varphi: T_n \to B$. Ist φ injektiv, so sind wir fertig, sei also $\operatorname{Ker}\varphi \neq 0$. Nach dem Scherungssatz und dem Weierstrass'schen Vorbereitungssatz dürfen wir ohne Einschränkung der Allgemeinheit annehmen, daß $\operatorname{Ker}\varphi$ ein Weierstrasspolynom $\omega \in k\langle X_1, \dots, X_{n-1}\rangle [X_n]$ in X_n enthält. Setzen wir $B' :=$ $= k\langle f_1, \dots, f_{n-1}\rangle$, so gilt für f_n eine ganze Gleichung

$$f_n^b + a_1 f_n^{b-1} + \dots + a_b = 0, \quad a_1, \dots, a_b \in B', \quad \text{wo} \quad \|a_1\| \leqslant 1, \dots, \|a_b\| \leqslant 1.$$

Mithin ist $\{1, f, \dots, f^{b-1}\}$ ein Erzeugendensystem des B'-Moduls B, d.h. die Injektion $B' \subsetneq B$ ist endlich. Zu B' gibt es nach Induktionsvoraussetzung ein $d \geq 0$ und einen endlichen k-Algebramonomorphismus $T_d \subsetneq B'$. Dann ist auch $T_d \subsetneq B$ endlich. Nach allgemeinen Sätzen der Dimensionstheorie ist d die Krullsche Dimension von B, w.z.b.w.

Bemerkung: Um eine bequeme Redeweise zu haben, nennen wir jeden endlichen Monomorphismus $T_d \subsetneq A$ eine *Realisierung von A über T_d*. Das Bild von T_d liegt abgeschlossen in B.

Wie in der affinen algebraischen Geometrie hat das Normalisierungslemma auch in der k-affinoiden Geometrie wichtige Konsequenzen. Unmittelbar folgt, da T_d erweiterungsendlich ist:

Folgerung 1: *Jede nullteilerfreie k-affinoide Algebra ist erweiterungsendlich.*

Dies impliziert z.B., daß die Normalisierung (=ganz-algebraischer Abschluß im vollen Quotientenring) einer reduzierten k-affinoiden Algebra wieder k-affinoid ist.

Folgerung 2: *Jede Restklassenalgebra $A/\mathfrak{m}$ einer k-affinoiden Algebra A nach einem maximalen Ideal $\mathfrak{m}$ ist ein endlich-algebraischer Erweiterungskörper des Grundkörpers k.*

Denn: Die Krullsche Dimension d eines Körper $A/\mathfrak{m}$ ist 0, also gibt es eine Realisierung von $A/\mathfrak{m}$ über $T_0 = k$, w.z.b.w.

3. k-affinoide Räume. – Die Elemente einer k-affinoiden Algebra sind analytische Funktionen. Wir bezeichnen mit $\bar{k}$ einen topologisch-algebraischen Abschluß von k und setzen:

$$\bar{E}^n := \{(c_1, \ldots, c_n) \in \bar{k}^n, |c_1| \leqslant 1, \ldots, |c_n| \leqslant 1\}.$$

$\mathfrak{O}$ sei die kohärente Garbe der $\bar{k}$-wertigen analytischen Funktionen auf $\bar{E}^n$.

Sei nun A eine k-affinoide Algebra, etwa $A = T_n/\mathfrak{a}$. Wir betrachten im $\bar{E}^n$ die Nullstellenmenge

$$X := N(\mathfrak{a}) := \{c \in \bar{E}^n, f(c) = 0 \text{ für alle } f \in \mathfrak{a}\}$$

des Ideals $\mathfrak{a}$. Die Idealgarbe $J := \mathfrak{a}\,\mathfrak{O}$ ist kohärent über $\bar{E}^n$. Der Hilbertsche Nullstellensatz ([6], Kap. III, § 3) impliziert:

X enthält Punkte mit über k algebraischen Koordinaten, es gilt $\mathfrak{a} = \Gamma(E^n, J) \cap T_n$.

Das Paar (X, H), wo $H := (\mathfrak{O}/J)|X$, ist ein analytischer Unterraum von $(\bar{E}^n, \mathfrak{O})$, man hat einen kanonischen Monomorphismus $A \hookrightarrow \Gamma(X, H)$. Das Tripel (X, H, A) heißt ein zu A gehörender *k-affinoider Raum*; es ist (bis auf Isomorphie) eindeutig durch A bestimmt. Die analytischen Funktionen aus A heißen *k-affinoide Funktionen.* Die Zuordnung $A \rightsquigarrow (X, H, A)$ ist ein Funktor, dabei sind die Morphismen der Kategorie der k-affinoiden Räume diejenigen analytischen Abbildungen, die k-affinoide Funktionen in k-affinoide Funktionen liften.

Das Normalisierungslemma besagt, daß es zu jedem k-affinoiden Raum (X, H, A) ein $d \geqslant 0$ und eine *endliche k-affinoide Abbildung* $\psi: (X, H, A) \rightarrow$

$(\bar{E}^d, \mathfrak{O}, T_d)$ mit $\psi(X)=\bar{E}^d$ gibt (analytische Überlagerungsabbildung), so daß A bzgl. $\psi^*: T_d \to A$ ein endlicher T_d-Modul ist.

Für jeden k-affinoiden Raum (X, H, A) ist die Teilmenge X_{alg} der *k-algebraischen Punkte von X* wohldefiniert, bzgl. einer Einbettung $(X, H, A) \hookrightarrow (\bar{E}^n, \mathfrak{O}, T_n)$ sind dies alle Punkte $(c_1, \ldots, c_n) \in X$ mit über k algebraischen Koordinaten. X_{alg} liegt dicht in X und bestimmt X eindeutig, die Funktionen $f \in A$ nehmen in den Punkten von X_{alg} stets k-algebraische Werte an.

Für jeden Punkt $p \in X_{alg}$ ist

$$\mathfrak{m}_p := \{f \in A, \quad f(p) = 0\}$$

ein *maximales Ideal in A*. Man hat somit eine natürliche Abbildung

$$\xi: X_{alg} \to Max\,A$$

von X_{alg} in die Menge $Max\,A$ aller maximalen Ideale von A. Diese Abbildung ist surjektiv; k-konjugierte Punkte haben dasselbe Bildideal.

TATE [*33*] benutzt als geometrischen Modellraum zu einer k-affinoiden Algebra A das Spektrum $Max\,A$, das in natürlicher Weise zu einem beringten Raum gemacht wird; in [*6*] wird der der naiven geometrischen Anschauung zugänglichere Raum X verwendet. $Max\,A$ ist kanonisch definiert; die Definition von (X, H, A) benutzt den nicht kanonisch gegebenen Körper $\bar{k}$. Beide Modelle liefern auf Grund der Abbildung $\xi: X_{alg} \to Max\,A$ dasselbe.

4. Spektralnorm. – Eine k-affinoide Algebra trägt a priori keine ausgezeichnete Norm, da verschiedene Restklassendarstellungen $T_n/\mathfrak{a}$ i.a. verschiedene Restklassennormen induzieren. Es gilt aber

Satz 2: *Jede reduzierte k-affinoide Algebra A ist* pm-*normierbar. Der Verzweigungsindex e der* pm-*Norm* $\| \ \|$ *von A (bzgl. k) ist endlich, speziell gilt:* $\|A\|^{e!} \subset |k|$.

Es genügt, den Satz für nullteilerfreie Algebren zu verifizieren. Sind nämlich $A_1, \ldots, A_s$ die Primkomponenten von A und $\| \ \|_1, \ldots, \| \ \|_s$ ihre pm-Normen, so wird der Summenring $\bigoplus_1^s A_i$ vermöge

$$\|(a_1, \ldots, a_s)\| = \max_{1 \leq i \leq s} \|a_i\|_i$$

pm-normiert. Da A abgeschlossen in $\bigoplus_1^s A_i$ liegt, erhält man durch Beschränkung eine die Topologie beschreibende pm-Norm auf A. Es gilt $e \leqslant e_1 + \cdots + e_s$, wenn e_i der Verzweigungsindex von A_i ist.

Im Falle nullteilerfreier Algebren gilt nun der präzisere:

Satz 2': *Es sei A eine nullteilerfreie k-affinoide Algebra und $T \subsetneq A$ eine Realisierung von A, es seien $\| \|_1, \ldots, \| \|_s$ die endlich vielen verschiedenen Fortsetzungen der kanonischen Bewertung von $Q(T)$ auf $Q(A)$. Dann ist die Maximumnorm*

$$\| \| = \max_{1 \leqslant i \leqslant s} \{\| \|_i\}$$

eine potenz-multiplikative T-Norm auf A, welche die Topologie von A beschreibt. Für jedes $f \in A$ gilt

$$\|f\| = \max_{1 \leqslant \nu \leqslant n} \sqrt[\nu]{\|a_\nu\|},$$

wenn $Y^n + a_1 Y^{n-1} + \cdots + a_n \in T[Y]$ das Minimalpolynom von f ist. Es gibt eine natürliche Zahl $e \leqslant [A:T]$, so daß gilt:

$$\|A\|^{e!} \subset |k|.$$

Beweis: Man zeigt zunächst, daß A eine normierte T-Algebra ist. Da $Q(T)$ nach Satz 4.4 quasi-vollkommen ist, ist A pm-normierbar nach § 2. Da A eine k-Banachalgebra ist, so ist diese pm-Norm nach § 2 die Maximumnorm $\max_{1 \leqslant i \leqslant s} \{\| \|_i\}$. Die letzte Behauptung folgt ebenfalls aus § 2, w.z.b.w.

Die pm-Norm einer reduzierten k-affinoiden Algebra A läßt sich geometrisch deuten. Jeder Körper $A/\mathfrak{m}$, $\mathfrak{m} \in Max\,A$, ist eine endliche Erweiterung von k; die Bewertung von k setzt sich eindeutig zu einer Bewertung $| \;|$ von $A/\mathfrak{m}$ fort. Für jedes $f \in A$ und jedes $\mathfrak{m} \in Max\,A$ bezeichne $f(\mathfrak{m}) \in A/\mathfrak{m}$ die Restklasse von f modulo $\mathfrak{m}$. Ist X der k-affinoide Raum A, so gilt $|f(\mathfrak{m})| = |f(p)|$, wenn $p \in X_{alg}$ ein ξ-Urbild von $\mathfrak{m}$ ist. Durch

$$\|f\|_{sp} := \sup_{\mathfrak{m} \in Max\,A} \{|f(\mathfrak{m})|\} = \sup_{p \in X} \{|f(p)|\}$$

wird jedem $f \in A$ eine nicht-negative reelle Zahl zugeordnet. Für jedes $c \in k$, $f \in A$ und jede natürliche Zahl $\nu \geqslant 1$ gilt offensichtlich:

$$\|cf^\nu\|_{sp} = |c| \cdot \|f\|_{sp}^\nu.$$

Tate ([*33*], theorem 5.1, p. 10) hat überdies gezeigt:

$f \in A$ ist topologisch-nilpotent genau dann, wenn $|f(\mathfrak{m})| < 1$ für alle $\mathfrak{m} \in Max\,A$ gilt.

Diese Informationen liefern nun schnell

Satz 3 ([5]): $\| \ \|_{sp}$ *ist für jede reduzierte k-affinoide Algebra die* pm-*Norm von A.*

Beweis: Sei $\| \ \|$ die pm-Norm von A. Da jede Restklassenabbildung $A \to A/\mathfrak{m}$ stetig ist und stetige (= beschränkte) Abbildungen zwischen pm-normierten Algebren Kontraktionen sind, folgt $|f(\mathfrak{m})| \leqslant \|f\|$ für alle $f \in A$ und alle $\mathfrak{m} \in Max A$. Mithin gilt jedenfalls: $\| \ \|_{sp} \leqslant \| \ \|$.

Gäbe es ein $g \in A$ mit $\|g\|_{sp} < \|g\|$, so kann man wegen $\|A\|^{e!} \subset |k|$ ein $c \in k$ finden, so daß für $f := cg^{e!} \in A$ gilt: $\|f\| = 1$. Dann folgt:

$$\|f\|_{sp} = |c| \cdot \|g\|_{sp}^{e!} < |c| \cdot \|g\|^{e!} = \|cg^{e!}\| = \|f\| = 1,$$

und f wäre nach TATE topologisch-nilpotent im Widerspruch zu $\|f\| = 1$. – Satz 3 ist bewiesen.

Auf Grund von Satz 3 und der Definition von $\| \ \|_{sp}$ nennen wir die pm-Norm einer reduzierten k-affinoiden Algebra A auch die *Spektralnorm* von A. Sie induziert auf dem zu A gehörenden analytischen Raum X die Topologie der gleichmässigen Konvergenz.

5. *Folgerungen aus Satz 3.* – Die Definitionsgleichung (1) der Spektralnorm liefert für nicht reduzierte k-affinoide Algebren A eine *Pseudonorm* auf A. Da jedes maximale Ideal von A das Nilradikal $\mathfrak{n}(A)$ enthält, so gilt

$$\|f\|_{sp} = \|\mathrm{red} f\|_{sp},$$

wenn $\mathrm{red}: A \to A/\mathfrak{n}(A)$ den natürlichen Epimorphismus bezeichnet. Daher folgt:

$f \in A$ ist genau dann nilpotent, wenn $f(\mathfrak{m}) = 0$ für alle $\mathfrak{m} \in Max A$ gilt.

Hierin ist speziell enthalten, daß jede k-affinoide Algebra ein *Jacobsonring* ist (d.h. jedes Primideal ist Durchschnitt maximaler Ideale).

Eine weitere wichtige Folgerung aus Satz 3 ist das

Maximumprinzip ([6], p. 456): *Ist A irgendeine k-affinoide Algebra, so gibt es zu jedem $f \in A$ ein $\mathfrak{m}_0 \in Max A$ mit $\|f\|_{sp} = |f(\mathfrak{m}_0)|$.*

Beweis: Wir dürfen A als reduziert voraussetzen. Angenommen, es gäbe ein $f \in A$, so daß $|f(\mathfrak{m})| < \|f\|_{sp}$ für alle $\mathfrak{m} \in Max A$ gilt. Wir dürfen $\|f\|_{sp} = 1$ annehmen. Dann wäre f aber, da $|f(\mathfrak{m})| < 1$ für alle $\mathfrak{m} \in Max A$ gilt, nach TATE topologisch-nilpotent, was, da $\| \ \|_{sp}$ eine pm-Norm ist, wegen $\|f\|_{sp} = 1$ unmöglich ist, w.z.b.w.

Ferner ergibt sich für reduzierte Algebren:

Jeder k-Algebrahomomorphismus $\varphi: A \to B$ zwischen reduzierten k-affi-

noiden Algebren ist eine Kontraktion:

$$\|\varphi(f)\|_{sp} \leqslant \|f\|_{sp} \quad \textit{für alle} \quad f \in A.$$

Angenommen, es gäbe ein $g \in A$ mit $\|\varphi(g)\|_{sp} > \|g\|_{sp}$. Dann gibt es ein $\mathfrak{m}_0 \in Max\,B$, so daß für $c := \varphi(g)(\mathfrak{m}_0) \in B/\mathfrak{m}_0$ gilt: $\|g\|_{sp} < |c|$. Wir dürfen $|c| = 1$ annehmen. Ist $Y^b + a_1 Y^{b-1} + \cdots + a_{b-1} Y + a_b \in k[Y]$ das Minimalpolynom von c über k, so folgt: $|a_\beta| \leqslant 1$, $\beta = 1, \ldots, b-1$, $|a_b| = 1$. Für

$$h := g^b + a_1 \cdot g^{b-1} + \cdots + a_{b-1} \cdot g \in A$$

gilt alsdann

$$\|h\| < 1 \quad \text{und} \quad \varphi(h)(\mathfrak{m}_0) = -a_b \in k.$$

Nun ist $a_b + h$ Einheit in A mit $a_b \sum (-1)^\nu \left(\frac{h}{a_b}\right)^\nu$ als Inversen (beachte: $|a_b| = 1$). Jedoch ist $\varphi(a_b + h) = a_b + \varphi(h)$ keine Einheit in B, da $\varphi(a_b + h)(\mathfrak{m}_0) = 0$. Widerspruch!

§ 6. Endlichkeit der Funktoren $A \rightsquigarrow \tilde{A}$ und $A \rightsquigarrow \mathring{A}$

1. Der Funktor $A \rightsquigarrow \tilde{A}$. – Ist A eine k-affinoide Algebra, so ist der Restklassenring

$$\tilde{A} = \mathring{A}/\mathfrak{t}(\mathring{A})$$

des Ringes der potenzbeschränkten Elemente von A nach dem Ideal $\mathfrak{t}(\mathring{A})$ der topologisch nilpotenten Elemente von A eine reduzierte Algebra über dem Restklassenkörper κ von k (vergl. § 2.2). Da jede k-affinoide Algebra A eine Realisierung über einer freien Algebra T_d besitzt, und da der Polynomring $\tilde{T}_d = \kappa[X_1, \ldots, X_d]$ noethersch und erweiterungsendlich ist, so folgt aus § 2.2:

Jede Algebra $\tilde{A}$ ist eine κ-affine Algebra, d.h. isomorph zu einer Restklassenalgebra eines freien Polynomringes $\kappa[X_1, \ldots, X_n]$ nach einem Ideal.

Die Algebra $\tilde{A}$ erbt wichtige algebraische Eigenschaften von A. So haben A und $\tilde{A}$ die *gleiche* (Krullsche) *Dimension*, weiter ist mit A auch $\tilde{A}$ *rein-dimensional*. Mit A ist auch A *zusammenhängend* (d.h. keine direkte Summe zweier Ideale $\mathfrak{a}_i \neq 0$, $i = 1, 2$). Indessen kann $\tilde{A}$ Nullteiler haben, wenn A nullteilerfrei ist, hier gilt:

Ist A reduziert, so ist $\tilde{A}$ genau dann nullteilerfrei, wenn die Spektralnorm von A eine Bewertung ist.

Die Algebren $\tilde{A}$ übernehmen in der nichtarchimedischen Funktionentheorie die Rolle des Tangentialraumes der lokalen komplexen Funktionentheorie. Dies beruht auf folgendem Endlichkeitssatz, dessen Analogon in der lokalen Theorie in [*15*], [*2*] bewiesen wurde.

Satz 1 (Endlichkeitssatz): *Ein k-Algebrahomomorphismus $\varphi: A \to B$ zwischen k-affinoiden Algebren A, B ist genau dann endlich, wenn der induzierte κ-Algebrahomomorphismus $\tilde{\varphi}: \tilde{A} \to \tilde{B}$ endlich ist.*

Daß die Endlichkeit von φ die Endlichkeit von $\tilde{\varphi}$ impliziert, wird in [7], p. 109 bewiesen. Die Umkehrung folgt aus [33], Prop. 4.2, wenn man beachtet, daß es stets die Topologien von A bzw. B definierende E-Algebren A_0 bezw. B_0 gibt, so daß gilt $B_0 = A_0 \langle t_1, \ldots, t_r \rangle$.

Die Abbildung $\tilde{\varphi}$ kann, wenn über A und B nichts weiter vorausgesetzt wird, surjektiv bezw. bijektiv sein, ohne daß dies für φ zutrifft. Für Homomorphismen zwischen freien Algebren gilt aber folgendes Analogon zum JACOBIschen Umkehrsatz ([6], p. 409):

Ein k-Algebrahomomorphismus $\varphi: T_m \to T_n$ ist bijektiv genau dann, wenn der zugehörige κ-Homomorphismus $\tilde{\varphi}: \tilde{T}_m \to \tilde{T}_n$ bijektiv ist.

Für $m = n = 1$ impliziert dies:

Ein k-Algebrahomomorphismus $\varphi: k\langle X \rangle \to k\langle Y \rangle$ ist genau dann ein Isomorphismus, wenn für die Potenzreihe $\varphi(X) = \sum_0^\infty a_\nu Y^\nu \in k\langle Y \rangle$ folgendes gilt:

$$|a_0| \leqslant 1, \quad |a_1| = 1, \quad |a_\nu| < 1 \quad \text{für alle} \quad \nu > 1.$$

Die Gruppe der k-affinoiden Automorphismen des Einheitskreises ist also *nicht* endlich-dimensional.

2. Der Funktor $A \rightsquigarrow \mathring{A}$. – Für viele Fragen der nichtarchimedischen Funktionentheorie ist der Funktor $A \rightsquigarrow \mathring{A}$ wichtiger als der Funktor $A \rightsquigarrow \tilde{A}$. Jedes maximale Ideal μ in $\mathring{A}$ umfaßt $\mathfrak{t}(\mathring{A})$, da jedes Element $1 - f \cdot \mathfrak{t}$, $f \in \mathring{A}$, $\mathfrak{t} \in \mathfrak{t}(\mathring{A})$, in $\mathring{A}$ das Inverse $\sum_0^\infty f^\nu \mathfrak{t}^\nu$ besitzt. Mithin ist jede κ-Algebra $\mathring{A}/\mu$ epimorphes Bild von $\tilde{A}$, also κ-affin und, da ein Körper, endlich-algebraisch über κ. Es gibt eine natürliche Abbildung $\alpha: Max\, A \to Max\, \mathring{A}$ des Spektrums der maximalen Ideale von A in das Spektrum der maximalen Ideale von $\mathring{A}$, die wie folgt beschrieben werden kann (Notationen wie im § 5, vgl. [33], p. 17). Für jedes $\mathfrak{m} \in Max\, A$ wird $\mathring{A}$ vermöge des Restklassenepimorphismus $\varepsilon: A \to A/\mathfrak{m}$ in den *Bewertungsring* $\widehat{A/\mathfrak{m}}^{\circ}$ des endlich-algebraischen Oberkörpers $A/\mathfrak{m}$ von k abgebildet. Das ε-Urbild des maximalen Ideals von $\widehat{A/\mathfrak{m}}^{\circ}$ ist ein maximales Ideal μ in $\mathring{A}$, das wir mit $\alpha(\mathfrak{m})$ bezeichnen. Es gilt

$$\alpha(\mathfrak{m}) = \{f \in \mathring{A} : |f(\mathfrak{m})| < 1\}.$$

TATE ([*33*], Theorem 6.4) hat bewiesen:

Satz 2: *Die Abbildung* $\alpha: Max\,A \to Max\,\mathring{A}$ *ist surjektiv.*

Dies hat wichtige Konsequenzen. So folgt z.B., daß die natürliche Abbildung des zu A gehörenden k-affinoiden Raumes X auf den zu $\tilde{A}$ gehörenden κ-affinen Raum $\tilde{X}$ surjektiv ist; weiter ergibt sich die für die Cohomologietheorie wichtige Aussage:

Sind $f_1, \ldots, f_s \in \mathring{A}$ *so beschaffen, daß zu jedem* $\mathfrak{m} \in Max\,A$ *ein Index* i, $1 \leqslant i \leqslant s$, *existiert mit* $|f_i(\mathfrak{m})| = 1$, *so gilt:*

$$\mathring{A} = \mathring{A} \cdot f_1 + \cdots + \mathring{A} f_s .$$

Schließlich impliziert Satz 2 noch, daß $\mathfrak{t}(\mathring{A})$ das *Jacobson-Radikal* von $\mathring{A}$ ist:

$$\mathfrak{t}(\mathring{A}) = \bigcap_{\mu \in Max\,\mathring{A}} \mu = \{ f \in \mathring{A} : 1\text{-}gf \text{ ist Einheit in } \mathring{A} \text{ für alle } g \in \mathring{A} \}.$$

Ist k diskret bewertet, so sind alle Ringe $\mathring{A}$ noethersch (da in diesem Falle alle Ringe $\mathring{T}_n$ noethersch und erweiterungsendlich sind), und jeder endliche k-Algebrahomomorphismus $\varphi: A \to B$ induziert einen endlichen E-Algebrahomomorphismus $\mathring{\varphi}: \mathring{A} \to \mathring{B}$, wenn B reduziert ist. Komplizierter liegen die Verhältnisse, wenn die Wertegruppe von k dicht in $\mathbb{R}^+$ liegt. Dann sind die Ringe $\mathring{A}$, $\mathring{B}$ nicht mehr noethersch und die Endlichkeit von $\varphi: A \to B$ impliziert nicht mehr einschränkungslos die Endlichkeit von $\mathring{\varphi}$. Es gilt vielmehr:

Satz 3 ([7], § 4): *Es sei k nicht diskret bewertet, es sei* $\varphi: T_d \subsetneq B$ *endlich, B sei reduziert. Dann ist* $\mathring{\varphi}: \mathring{T}_d \subsetneq \mathring{B}$ *genau dann endlich, wenn der Restklassengrad* $f = \dim_{Q(\tilde{T}_d)} Q(\tilde{B})$ *mit dem Grad* $n = \dim_{Q(T_d)} Q(B)$ *übereinstimmt:*

$$f = n .$$

Diese Gleichung ist nicht immer erfüllt, wenn $|k^*|$ dicht in $\mathbb{R}^+$ liegt. Ist z.B. $|k^*|$ nicht divisibel, so gibt es stets endlich-algebraische Erweiterungskörper k' von k, deren Verzweigungsindex e größer als 1 ist (man adjungiere z.B. zu k eine Nullstelle ξ des Polynoms $X^r - a$, wo $a \in E$, $a \neq 0$, $r > 1$, und $\sqrt[r]{|a|} \notin |k^*|$). k' ist, versehen mit der fortgesetzten Bewertung, eine 0-dimensionale k-affinoide Algebra. Da $ef \leqslant n$ für Körper stets richtig ist, folgt $f < n$; mithin ist die von der Injektion $k \subsetneq k'$ induzierte Abbildung $E \subsetneq \mathring{k}'$ nicht endlich, speziell ist E für solche Körper k nie erweiterungsendlich.

Unter zusätzlichen Annahmen über den Grundkörper k gilt stets die Gleichung $f = n$. So wird in [7] bewiesen:

Satz 4 (Endlichkeitssatz): *Ist k algebraisch abgeschlossen, so gilt immer*

$f=n$. Für solche Grundkörper k induziert jeder endliche k-Algebrahomomorphismus $\varphi: A \to B$ zwischen reduzierten k-affinoiden Algebren A, B einen endlichen E-Algebrahomomorphismus $\mathring{\varphi}: \mathring{A} \to \mathring{B}$, insbesondere ist also jeder Ring $\mathring{T}_n$, $n \geqslant 0$, erweiterungsendlich.

Literatur

[1] Artin, E., und B. L. van der Waerden: Die Erhaltung der Kettensätze der Idealtheorie bei beliebigen endlichen Körpererweiterungen. Nachr. Ges. d. Wiss. Göttingen, Math. Phys. Kl. 23–27 (1926).

[2] Bosch, S.: Endliche analytische Homomorphismen. Nachr. Akad. d. Wiss. Göttingen, Math. Phys. Kl., 41–49 (1967).

[3] Cartan, H.: Idéaux de fonctions analytiques de n variables complexes. Ann. Ecole Norm. Sup **61**, 149–197 (1944).

[4] Gerritzen, L.: Erweiterungsendliche Ringe in der nichtarchimedischen Funktionentheorie. Inv. Math. **2**, 178–190 (1966).

[5] Gerritzen, L.: Die Norm der gleichmässigen Konvergenz auf reduzierten k-affinoiden Algebren, ersch. in Journ. f.r.a. Math.

[6] Grauert, H. und R. Remmert: Nichtarchimedische Funktionentheorie. Weierstrass-Festschrift, Wissenschaftl. Abh. Arbeitsgem. f. Forsch. Nordrhein-Westfalen, **33**, 393–476 (1966).

[7] Grauert, H., und R. Remmert: Über die Methode der diskret bewerteten Ringe in der nichtarchimedischen Funktionentheorie. Inv. Math. **2**, 87–133 (1966).

[8] Grell, H.: Über die Gültigkeit der gewöhnlichen Idealtheorie in endlichen algebraischen Erweiterungen erster und zweiter Art. Math. Z. **40**, 503–505 (1936).

[9] Grothendieck, A.: Eléments de géométrie algébrique. IHES, Publ. Math. **20** (1964).

[10] Güntzer, U.: Zur Funktionentheorie einer Veränderlichen über einem vollständigen nichtarchimedischen Grundkörper. Arch. Math. **17**, 415–431 (1966).

[11] Güntzer, U.: Laurentreihen über vollständigen filtrierten Ringen. (Im Druck).

[12] Güntzer, U.: Modellringe in der nichtarchimedischen Funktionentheorie. Proc. Kon. Ned. Akad. v. Wetensch. A.**70**, 334–342 (1967).

[13] Hensel, K.: Über die arithmetischen Eigenschaften der Zahlen. Jahresber. DMV **16**, 299–319, 388–393, 473–496 (1907).

[14] Hensel, K.: Zahlentheorie. Berlin, Leipzig: Göschen, 1913.

[15] Houzel, Ch.: Géométrie analytique locale I. Sem. H. Cartan **13**, exp. 18 (1960/61).

[16] Kiehl, R.: Ausgezeichnete Ringe in der nichtarchimedischen analytischen Geometrie. Jour. f. reine und angew. Math. (1967).

[17] Krasner, M.: Essai d'une théorie des fonctions analytiques dans les corps valués complets. C.R. Acad. Sci. **222**, 37–40, 165–167 (1946).

[18] Lasker, E.: Zur Theorie der Moduln und Ideale. Math. Ann. **60**, 20–116 (1905).

[19] Lazard, M.: Les zéros des fonctions analytiques d'une variable sur un corps valué complet. IHES, Publ. Math. **14** (1962).

[20] Monna, A. F.: Linear Topological Spaces over Non-Archimedean Valued Fields. Proceedings of a Conference on Local Fields NUFFIC Summer School held at Driebergen in 1966, 56–65. Berlin – Heidelberg – New York: Springer 1967.

[21] Monna, A. F. und T. A. Springer: Sur la structure des espaces de Banach nonarchimediens. Proceed. Kon. Ned. Akad. v. Wetensch. A. **68**, 602–614 (1965).

[22] MORI, Y.: On the integral closure of an integral domain II. Bull. Kyoto Gakugei Univ. B **7**, 19–30 (1955).

[23] NAGATA, M.: Some remarks on local rings II. Mem. Coll. Sci. Univ. Kyoto, **28**, 109–120 (1953).

[24] NASTOLD, H. J.: Zur nichtarchimedischen Funktionentheorie. Math. Z. (1967).

[25] NOETHER, E.: Abstrakter Aufbau der Idealtheorie in algebraischen Zahl- und Funktionenkörpern. Math. Ann. **96**, 26–61 (1926).

[26] VAN DER PUT, M.: Algèbres de fonctions continues p-adiques. (Im Druck).

[27] RÜCKERT, W.: Zum Eliminationsproblem der Potenzreihenideale. Math. Ann. **107**, 250–281 (1933).

[28] SALMON, P.: Serie convergenti su un corpo non archimedeo con applicazione ai fasci analitici. Ann. Math. pur appl. (4), **65**, 113–126 (1964).

[29] SCHMIDT, F. K.: Über die Erhaltung der Kettensätze der Idealtheorie bei beliebigen endlichen Körpererweiterungen. Math. Zeitschr. **41**, 443–450 (1936).

[30] SCHÖBE, W.: Beiträge zur Funktionentheorie in nichtarchimedisch bewerteten Körpern. Universitas-Archiv, Math. Abt. Bd. 2 (= Bd. 42 d. Arch.). Münster/Westf.: Helios-Verlag 1930.

[31] SERRE, J. P.: Endomorphismes complètement continus des espaces de Banach p-adiques. IHES Publ. Math., **12**, 69–85 (1962).

[32] SPÄTH, H.: Der Weierstrass'sche Vorbereitungssatz. Journ. f. reine und angew. Math. **161**, 95–100 (1929).

[33] TATE, J.: Rigid analytic spaces. Private notes of J. Tate, reproduced with(out) his permission by IHES (1962).

[34] VAN DER WAERDEN, B. L.: Algebra II, 4. Aufl. Berlin-Göttingen-Heidelberg: Springer 1959.

[35] WASHNITZER, G. and MONSKY: Formal Cohomology. (Im Druck).

Sur les groupes de Galois attachés aux groupes p-divisibles [1]

J.-P. Serre

Introduction

Soit K un corps local de caractéristique 0 et de caractéristique résiduelle p, et soit C la complétion d'une clôture algébrique de K. Soit T le module de Tate ([*9*], n° 2.4) associé à un groupe p-divisible F, défini sur l'anneau des entiers de K. Tate a montré ([*9*], § 4, cor. 2 au th. 3) que $T \otimes C$ possède une décomposition analogue à celle de Hodge pour la cohomologie complexe. De nombreuses propriétés du module galoisien T sont implicitement contenues dans cette décomposition. Dans son séminaire au Collège de France, résumé dans [*8*], Tate en a indiqué un certain nombre (notamment lorsque l'action du groupe de Galois est abélienne). Dans ce qui suit, j'explicite une autre conséquence de cette décomposition de $T \otimes C$: si G est le groupe de Galois qui opère sur T, l'enveloppe algébrique de G contient (sous une hypothèse de semi-simplicité convenable) un groupe «de Mumford-Tate» p-adique (§ 3, ths. 1 et 2). Lorsqu'on fait certaines hypothèses supplémentaires sur F, on en déduit que G est *ouvert* dans $\mathrm{Aut}(T)$ (cf. § 5, th. 4). Ces hypothèses sont notamment vérifiées lorsque F est un groupe formel à 1 paramètre, n'admettant pas de multiplication complexe formelle (§ 5, th. 5).

§ 1. Enveloppes algébriques des groupes linéaires p-adiques

Soit V un espace vectoriel de dimension finie sur le corps p-adique $\mathbf{Q}_p$, et soit $\mathrm{Aut}(V)$ le groupe de ses automorphismes, muni de sa structure

[1] Le texte ci-dessous a été rédigé en décembre 1966. Il diffère de l'exposé oral par un plus grand usage de la théorie des groupes algébriques.

L'idée de remplacer les algèbres de Lie p-adiques par les groupes algébriques correspondants m'avait d'ailleurs été suggérée par Grothendieck il y a plusieurs années, en liaison avec sa théorie des «motifs».

naturelle de groupe de Lie p-adique. Soit G un sous-groupe compact de Aut(V). On sait (voir par exemple [7], p. 5.42) que G est un *sous-groupe de Lie* de Aut(V). Soit $\mathfrak{g}$ son algèbre de Lie; l'exponentielle définit un isomorphisme d'un voisinage ouvert de 0 dans $\mathfrak{g}$ (muni de la loi de groupe fournie par la formule de Hausdorff) sur un sous-groupe ouvert de G.

Proposition 1. – *Les conditions suivantes sont équivalentes:*

(a) *V est un G-module semi-simple.*

(b) *V est un $\mathfrak{g}$-module semi-simple.*

(c) $\mathfrak{g}$ *est une algèbre de Lie réductive* (i.e. produit d'une algèbre abélienne $\mathfrak{c}$ par une algèbre semi-simple $\mathfrak{s}$) *et V est un $\mathfrak{c}$-module semi-simple.*

Soit G_1 un sous-groupe ouvert distingué de G assez petit pour être contenu dans l'image de l'exponentielle. Tout sous-espace de V stable par $\mathfrak{g}$ l'est aussi par G_1, et inversement. Cela montre l'équivalence de (b) et de la condition suivante:

(a_1) V est un G_1-module semi-simple.

Mais G/G_1 est fini. On en déduit facilement (cf. Chevalley [*1*], tome III, p. 82, prop. 1) l'équivalence de (a) et (a_1). D'où (a)$\Leftrightarrow$(b). L'équivalence (b)$\Leftrightarrow$(c) est bien connue (Bourbaki, *Gr. et Alg. de Lie*, chap. I, § 6, n° 5, th. 4).

Nous supposerons à partir de maintenant que les conditions de la prop. 1 sont vérifiées. Nous allons associer à G un certain groupe algébrique réductif G_{alg}, de la manière suivante:

Soit d'abord $\mathbf{GL}_V$ le $\mathbf{Q}_p$-groupe algébrique des automorphismes de V. Cela signifie, par définition, que, si k est une $\mathbf{Q}_p$-algèbre commutative, le groupe $\mathbf{GL}_V(k)$ des points de $\mathbf{GL}_V$ à valeurs dans k est égal à Aut$(V \otimes k)$. En particulier, le groupe $\mathbf{GL}_V(\mathbf{Q}_p)$ s'identifie au groupe Aut(V) considéré plus haut.

Définition. – *On appelle enveloppe algébrique de G, et on note G_{alg}, le plus petit sous-groupe algébrique de $\mathbf{GL}_V$ dont le groupe des points contienne G.*

L'existence et l'unicité de G_{alg} sont immédiates (et ne nécessitent aucune hypothèse sur G). Si A est l'algèbre affine de $\mathbf{GL}_V$, l'idéal définissant G_{alg} est l'ensemble des $f \in A$ tels que $f(g)=0$ pour tout $g \in G$.

Utilisons maintenant l'hypothèse faite sur G. Décomposons l'algèbre de Lie $\mathfrak{g}$ en

$$\mathfrak{g} = \mathfrak{c} \times \mathfrak{s},$$

où $\mathfrak{c}$ est le centre de $\mathfrak{g}$, et $\mathfrak{s} = [\mathfrak{g}, \mathfrak{g}]$ est semi-simple. On sait (Chevalley

[*1*], tome II, p. 177, th. 15) que $\mathfrak{s}$ est *algébrique* (i.e. correspond à un groupe algébrique); si $\mathfrak{c}_{\mathrm{alg}}$ (resp. $\mathfrak{g}_{\mathrm{alg}}$) désigne la plus petite sous-algèbre de Lie algébrique de $\mathrm{End}(V)$ contenant $\mathfrak{c}$ (resp. contenant $\mathfrak{g}$), on a évidemment

$$\mathfrak{g}_{\mathrm{alg}} = \mathfrak{c}_{\mathrm{alg}} \times \mathfrak{s}.$$

L'algèbre $\mathfrak{c}_{\mathrm{alg}}$ est abélienne, et opère de façon semi-simple sur V. Il s'ensuit que $\mathfrak{g}_{\mathrm{alg}}$ est réductive. Le lien entre $\mathfrak{g}_{\mathrm{alg}}$ et G_{alg} est fourni par la proposition suivante:

Proposition 2. – (i) *L'algèbre de Lie de* G_{alg} *est* $\mathfrak{g}_{\mathrm{alg}}$.

(ii) G_{alg} *est réductif* (MUMFORD [*4*], p. 26, déf. 1.4).

(iii) *Toute composante connexe de* G_{alg} *rencontre* G.

Puisque $\mathfrak{g}_{\mathrm{alg}}$ est algébrique, il existe un sous-groupe algébrique connexe H^0 de $\mathbf{GL}_V$ d'algèbre de Lie $\mathfrak{g}_{\mathrm{alg}}$. Comme $\mathfrak{g}_{\mathrm{alg}}$ contient $\mathfrak{g}$, le groupe $H^0(\mathbf{Q}_p)$ contient un sous-groupe ouvert G_1 de G; on peut supposer G_1 distingué dans G. Soit (g_i) un ensemble fini de représentants des classes de G mod. G_1; la réunion H des $H^0 g_i$ est un sous-groupe algébrique de $\mathbf{GL}_V$ contenant G, et il est clair que c'est le plus petit possible. On a donc $H = G_{\mathrm{alg}}$, ce qui démontre (i) et (iii). L'assertion (ii) résulte de (i) (ou bien, si l'on préfère, du fait que V est un G_{alg}-module semi-simple et fidèle).

Corollaire. – *Si* $\mathfrak{c}$ *est algébrique, l'algèbre de Lie de* G_{alg} *est égale à* $\mathfrak{g}$, *et* G *est un sous-groupe ouvert de* $G_{\mathrm{alg}}(\mathbf{Q}_p)$.

La première assertion est évidente; la seconde en résulte puisque G et $G_{\mathrm{alg}}(\mathbf{Q}_p)$ sont des groupes de Lie p-adiques de même algèbre de Lie.

On peut donner une caractérisation des points de G_{alg} au moyen des *invariants* du groupe G. Plus précisément, soient r, s deux entiers $\geqslant 0$, et soit $T_{r,s}(V)$ le produit tensoriel de r copies de V et de s copies du dual V^* de V. Le groupe G opère (par transport de structure) sur $T_{r,s}(V)$; soit $T_{r,s}^0(V)$ le sous-espace de $T_{r,s}(V)$ formé des éléments invariants par G. Il est clair que G_{alg} laisse invariants les éléments de $T_{r,s}^0(V)$. De plus, cette propriété *caractérise* les points de G_{alg}:

Proposition 3. – *Soit* k *une* $\mathbf{Q}_p$*-algèbre commutative, et soit* $g \in \mathbf{GL}_V(k)$. *Pour que* g *appartienne à* $G_{\mathrm{alg}}(k)$, *il faut et il suffit que, pour tout couple* (r, s), *l'extension* $T_{r,s}(g)$ *de* g *à* $T_{r,s}(V \otimes k) = T_{r,s}(V) \otimes k$ *laisse invariants les éléments de* $T_{r,s}^0(V)$.

Notons d'abord qu'un élément de $T_{r,s}(V)$ est invariant par G_{alg} si et seulement si il appartient à $T_{r,s}^0(V)$. La prop. 3 est donc conséquence du lemme suivant:

Lemme 1. – *Soit* E *un corps de caractéristique* 0, *soit* V *un* E*-espace* *vectoriel de dimension finie, et soit H un sous-groupe algébrique réductif de* $\mathbf{GL}_V$. *Pour qu'un point de* $\mathbf{GL}_V$, *à valeurs dans une extension de E, soit un point de H, il faut et il suffit que, pour tout couple* (r, s), *il laisse invariants les éléments de* $T_{r,s}(V)$ *invariants par H.*

(En d'autres termes, un groupe réductif est déterminé par ses invariants tensoriels.)

Comme c'est là un résultat bien connu, je me bornerai à en esquisser la démonstration. On se ramène tout de suite au cas où le corps de base est algébriquement clos, ce qui permet d'identifier le groupe à l'ensemble de ses points rationnels; on peut aussi supposer que H est contenu dans le groupe unimodulaire $\mathbf{SL}_V$ (en effet, si $n = \dim. V$, on remplace V par $V \oplus \wedge^n V^*$). Soit alors g un élément de $\mathbf{GL}_V$, vérifiant la condition du lemme, et n'appartenant pas à H. Du fait que H est contenu dans $\mathbf{SL}_V$, H et Hg sont fermés dans $\mathrm{End}(V)$; puisque H est réductif, il existe donc une fonction polynôme f sur $\mathrm{End}(V)$, invariante par multiplication à gauche par H, et prenant les valeurs 0 sur H et 1 sur Hg (cf. [*4*], p. 29, cor. 1.2). Mais la représentation de H dans $\mathrm{End}(V)$ par multiplication à gauche est isomorphe à la somme directe de n copies de V. Comme f appartient à l'algèbre symétrique du dual de $\mathrm{End}(V)$, l'hypothèse faite sur g montre que f est invariante par g. Mais c'est absurde, puisque $f(1) = 0$ et $f(g) = 1$.

§ 2. Modules galoisiens du type de Hodge-Tate

A partir de maintenant, K désigne un corps muni d'une valuation discrète, et complet pour la topologie définie par cette valuation. On note A l'anneau des entiers de K, et $k = A/\mathfrak{m}$ le corps résiduel correspondant. On suppose K de caractéristique zéro, et k parfait de caractéristique p. On note $\bar{K}$ une clôture algébrique de K, et C sa complétion. Le groupe de Galois Gal_K de $\bar{K}/K$ opère sur C (cf. [9], § 3).

Soit $\mathbf{U}_p = \mathbf{Z}_p^*$ le groupe des unités de $\mathbf{Q}_p$, et soit

$$\chi : \mathrm{Gal}_K \to \mathbf{U}_p$$

le caractère de Gal_K donnant l'action de ce groupe sur les racines p^n-èmes de l'unité. Par définition, on a

$$sz = z^{\chi(s)}$$

pour tout $s \in \mathrm{Gal}_K$ et pour toute racine de l'unité z d'ordre une puissance de p.

Soit maintenant X un C-espace vectoriel de dimension finie muni d'une loi d'opération de Gal_K continue et semi-linéaire; on a

$$s(cx) = s(c)\,s(x) \quad \text{si} \quad s \in \mathrm{Gal}_K,\ c \in C,\ x \in X.$$

Si $i \in \mathbf{Z}$, notons X^i le sous-ensemble de X formé des éléments x tels que

$$sx = \chi(s)^i x \quad \text{pour tout} \quad s \in \mathrm{Gal}_K;$$

c'est un K-espace vectoriel. Posons $X(i) = X^i \otimes_K C$. L'injection $X^i \to X$ se prolonge en une application C-linéaire

$$\varepsilon_i : X(i) \to X.$$

TATE (séminaire au Collège de France) a démontré le résultat suivant:

Proposition 4. – *Soit* $\sum X(i)$ *la somme directe des* $X(i)$. *L'homomorphisme*

$$\varepsilon : \sum X(i) \to X,$$

somme des ε_i, *est injectif.*

Rappelons brièvement la démonstration. Soit (x_{ij}) une base de X^i sur K. Si ε n'était pas injectif, il existerait des $c_{ij} \in C$, non tous nuls, tels que

$$\sum c_{ij} x_{ij} = 0.$$

Parmi toutes les relations de ce genre, choisissons-en une de longueur minimum, et telle que $c_{i_0 j_0} = 1$ pour un couple (i_0, j_0) particulier. Si $s \in \mathrm{Gal}_K$, on a

$$\sum s(c_{ij})\, s(x_{ij}) = 0,$$

d'où

$$\sum s(c_{ij})\, \chi(s)^i x_{ij} = 0.$$

Utilisant la minimalité de la relation (c_{ij}), on en déduit que:

$$s(c_{ij})\, \chi(s)^i = \chi(s)^{i_0} c_{ij}, \quad \text{pour tout} \quad s \in \mathrm{Gal}_K.$$

Mais, d'après TATE [9], § 3, cette dernière relation entraîne $c_{ij} = 0$ pour $i \neq i_0$, et $c_{ij} \in K$ pour $i = i_0$, contrairement au fait que les (x_{ij}) sont linéairement indépendants sur K. D'où la proposition.

La proposition précédente permet *d'identifier* $\sum X(i)$ *à un sous-espace vectoriel de* X. Si ce sous-espace est égal à X tout entier, nous dirons que X est *du type Hodge-Tate*, ou admet une décomposition de type (HT).

Remarque. Si X est de type (HT), il en est de même de son dual X^*, et l'on a:

$$X^*(i) = X(-i)^*;$$

la vérification est immédiate. De même, tout produit tensoriel de modules galoisiens de type (HT) est de type (HT).

§ 3. Le groupe de Mumford-Tate p-adique

Les notations étant celles du § 2, soit V un $\mathbf{Q}_p$-espace vectoriel de dimension finie muni d'une loi d'opération de Gal_K continue et linéaire. Soit G l'image de Gal_K dans $\mathrm{Aut}(V)$; c'est un sous-groupe compact de $\mathrm{Aut}(V)$.

Faisons les deux hypothèses suivantes:

(H.1) – *V est un G-module semi-simple.*

(H.2) – *Le module galoisien $V_C = V \otimes_{\mathbf{Q}_p} C$ est du type Hodge-Tate.*

(Précisons que Gal_K opère sur V_C par transport de structure, i.e. par la formule $s(v \otimes c) = s(v) \otimes s(c)$.)

L'hypothèse (H.1) permet d'appliquer à G les définitions et résultats du § 1; en particulier, le groupe G_{alg} est défini. On notera que $G_{\mathrm{alg}}(C)$ est un sous-groupe du groupe $\mathrm{Aut}(V_C)$ des automorphismes C-linéaires de V_C.

Soit d'autre part $\lambda \in C^*$; notons $\varphi(\lambda)$ l'automorphisme de V_C qui est l'homothétie de rapport λ^i sur la i-ème composante $V_C(i)$ de V_C (au sens du § 2). On définit ainsi un homomorphisme $\varphi: C^* \to \mathrm{Aut}(V_C)$. Soit $\Phi = \mathrm{Im}(\varphi)$; c'est un sous-groupe de $\mathrm{Aut}(V_C)$.

Théorème 1. – *Le groupe Φ est contenu dans $G_{\mathrm{alg}}(C)$.*

Soit (r, s) un couple d'entiers $\geqslant 0$, et soit $W = T_{r,s}(V)$. Vu la prop. 3, il suffit de prouver que tout élément de W invariant par Gal_K est invariant par Φ (lorsqu'on l'identifie à un élément de $T_{r,s}(V_C) = W_C$). Or:

(a) *W_C est du type Hodge-Tate.*

(b) *Si $\lambda \in C^*$, l'action de $\varphi(\lambda)$ sur W_C est donnée par:*

$$\varphi(\lambda)\, x = \lambda^i x \quad \textit{si} \quad x \in W_C(i).$$

En effet, (a) et (b) sont vrais pour $W = V$ (i.e. $r=1$, $s=0$), et, si elles sont vraies pour W_1 et W_2, elles le sont aussi pour W_1^* et $W_1 \otimes W_2$, en vertu de la remarque de la fin du § 2.

Ceci étant, si $w \in W$ est invariant par Gal_K, w appartient *a fortiori* à W_C^0, donc aussi à $W_C(0) = W_C^0 \otimes_K C$, et (b) montre que $\varphi(\lambda)\, w = w$ pour tout $\lambda \in C^*$. L'élément w est donc bien invariant par Φ, cqfd.

Soit M le plus petit sous-groupe algébrique de $\mathbf{GL}_V$ (au sens strict du terme, i.e. «défini sur $\mathbf{Q}_p$») dont le groupe des points à valeurs dans C contienne Φ. Je dirai que M est *le groupe de Mumford-Tate* du module galoisien V (c'est en effet l'analogue p-adique de celui défini dans [5]). Le théorème 1 est visiblement équivalent au suivant:

Théorème 2. – *Le groupe G_{alg} contient le groupe M.*

Remarques. 1) Le groupe Φ est un C-sous-groupe algébrique connexe de $\mathbf{GL}_V(C)$. Il en résulte que M est connexe, donc *contenu dans la composante neutre* de G_{alg}.

2) Soit $d\varphi$ l'endomorphisme de V_C qui est l'homothétie de rapport i sur $V_C(i)$. Les théorèmes 1 et 2, traduits en termes d'algèbres de Lie, signifient que:

$$d\varphi \in \mathfrak{g}_{\mathrm{alg}} \otimes C.$$

3) Soit K_{nr} l'extension non ramifiée maximale de K contenue dans $\bar{K}$. L'image I de $\mathrm{Gal}(\bar{K}/K_{nr})$ dans G est le *sous-groupe d'inertie* de G. Comme I est distingué dans G, V est un I-module semi-simple et le groupe I_{alg} est défini. *On a* $M \subset I_{\mathrm{alg}}$; cela se voit en appliquant le th. 2 sur le corps de base $\hat{K}_{nr}$ complété de K_{nr}. (On n'aurait donc rien perdu si l'on avait supposé le corps résiduel k algébriquement clos.)

§ 4. Un cas particulier

Conservons les notations des §§ 2, 3, et faisons sur V les hypothèses suivantes:

(H*.1) *V est un $\mathfrak{g}$-module absolument simple.*

(H*.2) *V_C est somme directe de $V_C(0)$ et $V_C(1)$.*

(H*.3) *Les dimensions n_0 et n_1 de $V_C(0)$ et $V_C(1)$ sont $\geqslant 1$ et premières entre elles.*

Remarques. – 1) Les hypothèses (H*.1) et (H*.2) entraînent évidemment les hypothèses (H.1) et (H.2) du § 3.

2) (H*.1) équivaut à dire que le $\mathfrak{g}$-module V est *semi-simple* et que *son commutant est réduit aux homothéties* (BOURBAKI, *Alg.* VIII, § 13, n° 4, cor. à la prop. 5). En particulier, si $\mathfrak{c}$ désigne le centre de $\mathfrak{g}$, on voit que $\mathfrak{c}$ est, ou bien réduit à 0, ou bien égal à l'ensemble des homothéties de V. Dans les deux cas, c'est une algèbre de Lie algébrique, et l'on a donc $\mathfrak{g}_{\mathrm{alg}} = \mathfrak{g}$.

Théorème 3. – *Sous les hypothèses ci-dessus, on a* $G_{\mathrm{alg}} = \mathbf{GL}_V$.

Vu la remarque 2) ci-dessus, cet énoncé équivaut à:

Corollaire 1. – *L'algèbre de Lie $\mathfrak{g}$ de G est égale à* $\mathrm{End}(V)$.

Il équivaut aussi à:

Corollaire 2. – *Le groupe G est un sous-groupe* ouvert *de* $\mathrm{Aut}(V)$.

(En d'autres termes, l'action de Gal_K sur V est aussi peu triviale que possible.)

Démonstration. – Posons, pour simplifier $E = V_C$, $E_0 = V_C(0)$ et

$E_1 = V_C(1)$. Soit H la composante neutre du C-groupe algébrique $G_{\mathrm{alg}}(C)$ (du fait que C est algébriquement clos, nous nous permettons d'identifier un C-groupe algébrique à l'ensemble de ses points rationnels). Tout revient à montrer que $H = \mathbf{GL}_E$. Or, H jouit des trois propriétés suivantes:

(a) *H est un sous-groupe réductif connexe de $\mathbf{GL}_E$, de commutant réduit aux homothéties.*

(b) *H contient le groupe Φ formé des automorphismes de E qui sont l'identité sur E_0 et une homothétie sur E_1.*

Cela résulte du théorème 1.

(c) *Les dimensions n_0 et n_1 de E_0 et E_1 sont premières entre elles.*

Le théorème 3 est donc une conséquence du résultat suivant, qui est un pur énoncé de théorie des groupes algébriques:

Proposition 5. – *Tout sous-groupe algébrique H de $\mathbf{GL}_E$ vérifiant les conditions* (a), (b), (c) *ci-dessus est égal à $\mathbf{GL}_E$.*

La démonstration comporte plusieurs étapes:

1) Soit T (resp. S) la composante neutre du centre de H (resp. son groupe des commutateurs). On a $H = T \cdot S$, $T \cap S$ est fini, et S est semi-simple. Vu (a), T est, soit réduit à $\{1\}$, soit égal au groupe $\mathbf{G}_m$ des homothéties. Dans le premier cas, $H = S$ serait contenu dans le groupe unimodulaire $\mathbf{SL}_E$, ce qui est absurde car les éléments de Φ ne sont pas tous de déterminant 1. On a donc $T = \mathbf{G}_m$ et il va nous suffire de prouver que $S = \mathbf{SL}_E$.

2) Soit Θ le tore de dimension 2 formé des automorphismes de E qui sont égaux à une homothétie sur E_0 et à une autre homothétie sur E_1. On a $\Theta = \mathbf{G}_m \cdot \Phi$, d'où $\Theta \subset H$ d'après ce qui précède. Soit $\psi : C^* \to \Theta$ l'homomorphisme défini par:

$$\psi(\lambda)\, x = \lambda^{n_1} x \quad \text{si} \quad x \in E_0$$
$$\psi(\lambda)\, x = \lambda^{-n_0} x \quad \text{si} \quad x \in E_1 .$$

On a $\det \psi(\lambda) = \lambda^{n_0 n_1 - n_1 n_0} = 1$ pour tout $\lambda \in C^*$. L'image Ψ de ψ est donc un sous-groupe connexe de $H \cap \mathbf{SL}_E$, donc aussi *un sous-groupe* de S.

3) Montrons maintenant que S est *simple* (i.e. tout sous-groupe algébrique distingué de S est, soit fini, soit égal à S). Sinon, en effet, on aurait $S = (S' \times S'')/N$, avec S', S'' semi-simples non réduits à $\{1\}$, et N sous-groupe fini du centre de $S' \times S''$. Le $(S' \times S'')$-module E peut s'écrire comme produit tensoriel:

$$E = E' \otimes E'' ,$$

où E' (resp. E'') est un S'-module (resp. S''-module) absolument simple. Soit $\tilde{\Psi}$ la composante neutre de l'image réciproque de Ψ dans $S' \times S''$. C'est un tore de dimension 1, donc isomorphe au groupe $\mathbf{G}_m$. Choisissons un isomorphisme $\sigma: \tilde{\Psi} \to \mathbf{G}_m$. Notons respectivement f, f_0, f_1, f', f'' les caractères des représentations E, E_0, E_1, E', E'' de $\tilde{\Psi}$. Ce sont des polynômes à coefficients entiers positifs en σ et σ^{-1}. On a

$$f' \cdot f'' = f = f_0 + f_1$$

et

$$f_0 = n_0 \sigma^a, \quad f_1 = n_1 \sigma^b, \quad \text{avec} \quad a, b \in \mathbf{Z}.$$

On a $a \neq b$; sinon, en effet, Ψ opèrerait sur E par homothéties. De plus:

$$f = n_0 \sigma^a + n_1 \sigma^b.$$

Mais on vérifie tout de suite qu'un tel binôme ne peut se décomposer en produit de deux polynômes à coefficients entiers $\geqslant 0$ que de façon triviale, l'un des facteurs étant un monôme $c\sigma^d$. L'entier c doit diviser n_0 et n_1, donc doit être égal à 1 vu l'hypothèse $\operatorname{pgcd}(n_0, n_1) = 1$. On a donc, par exemple, $f' = \sigma^d$, d'où $\dim. E' = 1$. Mais c'est absurde, car l'image de S' dans S opèrerait trivialement sur E, contrairement au fait que E est un S-module fidèle. Le groupe S est donc bien un groupe *simple*.

4) Soit $h = n_0 + n_1$ la dimension de E, et soit μ_h le groupe des racines h-èmes de l'unité. On a $\lambda^{n_1} = \lambda^{-n_0}$ si $\lambda \in \mu_h$. L'homomorphisme ψ de 2) transforme donc $\lambda \in \mu_h$ en une *homothétie*. Comme n_0 et n_1 sont premiers entre eux, ψ est injectif. On en conclut finalement que Ψ, donc *a fortiori* S, *contient le groupe* μ_h, identifié à un sous-groupe du groupe des homothéties. En particulier, μ_h *est contenu dans le centre de* S.

5) Vu 3) et 4), il ne nous reste plus qu'à démontrer le lemme suivant:

Lemme 2. – *Soit S un groupe algébrique simple, et soit $\varrho: S \to \mathbf{GL}_E$ une représentation linéaire non triviale de S, de dimension h. Si le centre de S est d'ordre multiple de h, ϱ est un isomorphisme de S sur $\mathbf{SL}_E$.*

Distinguons deux cas:

a) Le groupe S est de type $\mathbf{A}_n$ (au sens de la classification des groupes algébriques simples), donc quotient de $\mathbf{SL}(n+1)$ par un sous-groupe de son centre. Par hypothèse, h divise $n+1$. D'où $\dim. S \geqslant \dim. \mathbf{SL}_E$, et, comme le noyau de ϱ est fini, on en conclut d'abord que ϱ applique S sur $\mathbf{SL}_E$, puis que c'est un isomorphisme ($\mathbf{SL}_E$ étant simplement connexe).

b) Le groupe S n'est pas de type $\mathbf{A}_n$. La classification des groupes simples montre alors que le centre de S a au plus 4 éléments. D'où $h \leqslant 4$. Le cas $h = 1$ est impossible. Le cas $h = 2$ donnerait $S = \mathbf{SL}(2)$, qui est exclu.

Le cas $h=3$ n'est possible que si S est de type $\mathbf{E}_6$; mais la dimension de $\mathbf{E}_6$ est bien trop grande pour que ce groupe puisse être plongé dans $\mathbf{SL}(3)$. De même, le cas $h=4$ n'est possible que pour un groupe de type $\mathbf{D}_n$, avec $n \geqslant 4$ (pour $n \leqslant 3$, $\mathbf{D}_n$ est, soit isomorphe à $\mathbf{A}_n$, soit non simple); comme la dimension de $\mathbf{D}_4$ est strictement supérieure à celle de $\mathbf{SL}(4)$, on conclut comme précédemment.

Ceci achève la démonstration du lemme 2, et, avec elle, celles de la prop. 5 et du th. 3.

Remarque. Même lorsqu'on ne fait plus l'hypothèse (H*.3), l'existence dans S d'un tore Ψ de dimension 1, ayant pour caractère un binôme $n_0\sigma^a + n_1\sigma^b$, peut être utilisée pour déterminer (au moins en partie) la structure de S.

§ 5. Application aux groupes p-divisibles

Conservons les notations des paragraphes précédents. Soit F un *groupe p-divisible de hauteur h* sur l'anneau A des entiers de K (pour tout ce qui concerne les groupes p-divisibles, voir [9]). Soit T le module de Tate de F, et soit $V = T \otimes \mathbf{Q}_p$. On sait ([9], n° 2.4) que dim. $V = h$, et que Gal_K opère continûment sur T et sur V.

Nous aurons besoin de deux des principaux résultats de la théorie de Tate. Tout d'abord la décomposition de V_C ([9], § 4, cor. 2 au th. 3):

Proposition 6. – *$V_C = V \otimes C$ possède une décomposition de Hodge-Tate de la forme:*

$$V_C = V_C(0) \oplus V_C(1).$$

De plus, la dimension n_1 (resp. n_0) de $V_C(1)$ (resp. de $V_C(0)$) est égale à la dimension de la composante neutre de F (resp. de son dual F').

(Rappelons, cf. [9], n° 2.2, que la composante neutre F^0 de F peut être identifiée à un *groupe formel* sur A, au sens de Dieudonné, Lazard, Lubin; par la «dimension» de F^0, on entend le «nombre de paramètres» du groupe formel en question.)

Si K' est une extension finie de K, d'anneau de valuation A', notons $\mathrm{End}_{A'}(F)$ l'anneau des endomorphismes du groupe p-divisible $F \times_A A'$ déduit de F par extension des scalaires de A à A'. Soit $\mathrm{End}(F)$ la limite inductive (qui est en fait une réunion) des $\mathrm{End}_{A'}(F)$ lorsque K' parcourt toutes les sous-extensions finies de $\bar{K}$ (cf. Lubin [2] pour le cas de dimension 1).

Proposition 7. – (a) *Le commutant de G dans T est égal à* $\mathrm{End}_A(F)$.

(b) *Le commutant de G dans V est égal à* $\mathrm{End}_A(F)\otimes\mathbf{Q}_p$.

(c) *Le commutant de* $\mathfrak{g}$ *dans V est égal à* $\mathrm{End}(F)\otimes\mathbf{Q}_p$.

L'assertion (a) est démontrée dans [9], § 4, cor. 1 au th. 4; (b) résulte de (a) par produit tensoriel avec $\mathbf{Q}_p$. Enfin, (c) résulte de (b) et du fait que le commutant de $\mathfrak{g}$ dans V est égal à la réunion des commutants des sous-groupes ouverts de G.

Nous pouvons maintenant appliquer le théorème 3. On en tire:

Théorème 4. – *Faisons sur F les hypothèses suivantes:*

(a) *V est un G-module semi-simple.*

(b) $\mathrm{End}(F)=\mathbf{Z}_p$.

(c) *Les dimensions n_1 et n_0 de F et de son groupe dual F' sont $\geqslant 1$ et premières entre elles.*

On a alors $G_{\mathrm{alg}}=\mathbf{GL}_V$, $\mathfrak{g}=\mathrm{End}(V)$ *et G est un sous-groupe ouvert de* $\mathrm{Aut}(V)$.

En effet, (H*.1) résulte de (a) et (b), compte tenu de la prop. 7, et (H*.2) et (H*.3) résultent de la prop. 6 et de (c).

Remarque. Les hypothèses (a), (b), (c) entraînent que F est connexe (c'est donc un *groupe formel*). En effet, si F^0 désigne la composante neutre de F, $V(F^0)$ est un sous-espace de $V(F)=V$ stable par G. Vu (a) et (b), on a donc $V(F^0)=0$ ou $V(F^0)=V$. Le premier cas entraîne $F^0=0$, d'où $n_1=0$, ce qui est exclu d'après (c). Le second cas entraîne $F=F^0$. De même, le dual de F est un groupe formel.

Nous allons maintenant appliquer le théorème 4 aux *groupes formels de dimension* 1 (au sens de LUBIN [2], [3]), i.e. au cas $n_1=1$. On a tout d'abord:

Proposition 8. – *Supposons F connexe et de dimension* 1. *Pour tout $z\in V$, $z\neq 0$, on a $\mathfrak{g}\cdot z=V$. En particulier, V est un $\mathfrak{g}$-module simple.*

Par hypothèse, l'algèbre affine de F est isomorphe à l'algèbre des séries formelles $A[[X]]$. Choisissons un tel isomorphisme (ce qui revient à considérer F comme un groupe formel). La multiplication par p dans F est donnée par une série formelle

$$f(X)=\sum_{n=1}^{\infty} a_n X^n, \qquad a_n\in A, \quad a_1=p.$$

Puisque F est de hauteur h, on a

$$a_n\in\mathfrak{m} \quad \text{pour} \quad n<p^h, \quad \text{et} \quad a_n\notin\mathfrak{m} \quad \text{pour} \quad n=p^h,$$

cf. LUBIN [2].

Soit $\bar{A}$ (resp. $\bar{\mathfrak{m}}$) l'anneau (resp. l'idéal) de valuation de $\bar{K}$. Pour tout entier $n \geqslant 0$, notons $f^{(n)}$ le n-ème itéré de f (i.e. *la multiplication par* p^n dans F), et soit T_n l'ensemble des $x \in \bar{\mathfrak{m}}$ tels que $f^{(n)}(x)=0$. On sait (cf. par exemple [3]) que T_n est un groupe (pour la loi de groupe formel définissant F), et que ce groupe est isomorphe à $(\mathbf{Z}/p^n\mathbf{Z})^h$; on a $T = \varprojlim T_n$. Notons T_n' l'ensemble des éléments de T_n d'ordre p^n; on a

$$T_n' = T_n - T_{n-1}.$$

Lemme 3. – *Il existe un nombre $c>0$ tel que, pour tout $n \geqslant 1$, et tout $x \in T_n'$, l'indice de ramification de l'extension $K(x)/K$ soit $\geqslant c \cdot p^{nh}$.*

Admettons provisoirement ce lemme, et démontrons la prop. 8. Quitte à transformer z par une homothétie, on peut supposer que $z \in T$ et $z \notin pT$. Pour tout $n \geqslant 1$, l'image z_n de z dans $T/p^nT = T_n$ appartient à T_n'. Soit d'autre part $U = G \cdot z$ *l'orbite* de z par G. C'est une *sous-variété analytique p-adique* de T et son espace tangent au point z est égal à $\mathfrak{g} \cdot z$ (cf. par exemple [7], LG, p. 4.12). D'autre part, l'image U_n de U dans T/p^nT est égale à $G \cdot z_n$, ensemble des conjugués de z_n. Mais, d'après le lemme 3, on a $\mathrm{Card}(G \cdot z_n) = [K(z_n):K] \geqslant c \cdot p^{nh}$. D'où

$$\mathrm{Card}(U_n) \geqslant c \cdot p^{nh}.$$

Soit alors μ la mesure de Haar de T, normalisée de telle sorte que $\mu(T)=1$. La formule ci-dessus entraîne que $\mu(U) \geqslant c$. Mais il est immédiat qu'une sous-variété analytique p-adique de T n'a une mesure >0 que si son intérieur est non vide. Comme U est une orbite, on en conclut que U est *ouvert* dans T. Son espace tangent en z est donc V tout entier. D'où $\mathfrak{g} \cdot z = V$.

Reste à démontrer le lemme 3. Soit v la valuation de $\bar{K}$, normalisée de telle sorte que $v(p)=1$. Soit c_1 la borne inférieure des $v(a_n)$, pour $n < p^h$. D'après ce qui a été dit plus haut, on a $c_1 > 0$. Soit $\varphi: \mathbf{R}_+ \to \mathbf{R}_+$ l'application donnée par la formule

$$\varphi(\alpha) = \mathrm{Inf}\,(p^h \alpha, \alpha + c_1).$$

C'est une bijection strictement croissante de $\mathbf{R}_+$ sur $\mathbf{R}_+$; soit ψ la bijection réciproque. Si $x \in \bar{\mathfrak{m}}$, la formule

$$f(x) = \sum a_n x^n$$

montre que

$$v(f(x)) \geqslant \varphi(v(x)),$$

ou encore

$$v(x) \leqslant \psi(v(f(x))).$$

De plus, pour $v(x) > c_2 = c_1/(p^h - 1)$, la valuation du terme correspondant à $n = p^h$ est strictement inférieure à celle des autres termes, et l'on a donc

$$v(f(x)) = p^h v(x).$$

Soit d_n la borne supérieure des $v(x)$, pour $x \in T'_n$. Les d_n sont >0. Si $x \in T'_n$, on a $f(x) \in T'_{n-1}$; d'où:

$$d_n \leqslant \psi(d_{n-1}) \quad \text{et} \quad d_n \leqslant \psi^{(n-1)}(d_1).$$

Mais, pour tout $\alpha > 0$, les itérés $\psi^{(n)}(\alpha)$ tendent vers 0. Il existe donc un entier n_0 tel que, si $n \geqslant n_0$, on ait $d_n < c_2$, d'où $d_{n+1} = d_n/p^h$. On obtient ainsi l'existence d'une constante $c_3 > 0$ telle que:

$$v(x) \leqslant c_3/p^{nh} \quad \text{si} \quad x \in T'_n.$$

Soit c_4 la valuation d'une uniformisante de K, et soit e l'indice de ramification de $K(x)/K$. On a évidemment:

$$v(x) \geqslant c_4/e.$$

En comparant, on trouve $e \geqslant c \cdot p^{nh}$, avec $c = c_4/c_3$, ce qui démontre le lemme.

On voit ainsi que la condition (a) du théorème 4 est automatiquement vérifiée en dimension 1. La condition (c) l'est aussi pourvu que $h \geqslant 2$, puisque $n_1 = 1$ et $n_0 = h - 1$. Comme le cas $h = 1$ est trivial, on en déduit finalement:

Théorème 5. – *Soit F un groupe formel défini sur A, de dimension* 1 *et de hauteur finie h. Supposons que* $\mathrm{End}(F) = \mathbf{Z}_p$. *L'image G de* Gal_K *dans* $\mathrm{Aut}(V)$ *est ouverte.*

(L'hypothèse $\mathrm{End}(F) = \mathbf{Z}_p$ signifie que F n'a pas de «multiplication complexe formelle», au sens de LUBIN [*2*].)

Remarques. – 1) Lorsqu'on ne fait plus l'hypothèse que $\mathrm{End}(F) = \mathbf{Z}_p$, on peut quand même déterminer l'algèbre de Lie $\mathfrak{g}$ de G. On trouve que c'est le *commutant* de $\mathrm{End}(F)$ dans V. La démonstration est analogue à celle du cas particulier traité ici, mais sensiblement plus compliquée.

2) Les résultats ci-dessus s'appliquent au *groupe d'inertie I* de G; cela se voit en passant au corps $\hat{K}_{nr}$, et en remarquant que $\mathrm{End}(F)$ ne change pas par cette opération, en vertu de Lubin [*2*], § 2 (c'est là une propriété

spéciale aux groupes de dimension 1). On en déduit que *I est un sous-groupe ouvert de G*; je ne connais pas de démonstration directe de ce fait.

3) Le groupe I possède deux filtrations naturelles v et w, définies de la manière suivante:

$$v(g) \geqslant n \Leftrightarrow (g-1)(T) \subset p^n T,$$
$$w(g) \geqslant n \Leftrightarrow g \in I^n$$

(I^n désigne le n-ème groupe de ramification de I, dans la notation supérieure, cf. [6], chap. IV, § 3).

Quelle relation y a-t-il entre v et w? Si e_K désigne l'indice de ramification absolu de K, est-il vrai que l'on a

$$w = e_K \cdot v + O(1)?$$

Une question analogue se pose pour tous les groupes de Lie p-adiques qui sont des groupes de Galois.

Bibliographie

[1] CHEVALLEY, C.: Théorie des groupes de LIE, II et III. Publ. Inst. Math. Univ. Nancago, Paris: Hermann 1951 et 1955.

[2] LUBIN, J.: One-parameter formal LIE groups over p-adic integer rings. Ann. of Maths. **80**, 464–484 (1964).

[3] LUBIN, J.: Finite subgroups and isogenies of one-parameter formal LIE groups. Ann. of Maths. **85**, 296–302 (1961).

[4] MUMFORD, D.: Geometric invariant theory. Ergeb. der Math., Neue Folge, Bd. 34. Berlin, Heidelberg, New York, Springer 1965.

[5] MUMFORD, D.: Families of abelian varieties. Algebraic groups and discontinuous subgroups. Proc. Symp. Pure Maths. **9**, A.M.S. (1966).

[6] SERRE, J-P.: Corps Locaux. Publ. Inst. Math. Univ. Nancago, VIII, Paris: Hermann 1962.

[7] SERRE, J-P.: LIE algebras and LIE groups. New York: Benjamin 1965.

[8] SERRE, J-P.: Résumé des cours 1965–66. Annuaire du Collège de France, 49–58, (1966–67).

[9] TATE, J.: p-divisible groups, ce volume p. 158–183.

The Conjectures of Birch and Swinnerton-Dyer, and of Tate

P. SWINNERTON-DYER

1. Introduction

In the last few years, it has become increasingly evident that the study of the zeta-function of an algebraic variety can yield valuable information about that variety, some of which cannot easily be obtained in any other way. Most of this information is number-theoretic – that is to say, it refers to objects which depend on the ground field and which are only of interest when the ground field is finitely generated. But some of it refers to objects, such as the Néron-Severi group, which are also of interest to classical algebraic geometers – and about which classical algebraic geometry has little to say.

The first indication of this should have been the pre-war work of SIEGEL on quadratic forms. Indeed Siegel himself wrote of that work that he hoped it would rescue the zeta-function from the neglect into which it was falling. But the influence of fashion, and the deeper and apparently more elegant reformulation of Siegel's work by TAMAGAWA in terms of measure theory, led to a general belief that zeta-functions were not essentially involved in Siegel's results.

Siegel's theorems were rigorously proved. Most of the subsequent work is conjecture, based on the examination of special cases and on ex post facto heuristic arguments which have been adequately described elsewhere. The next step was the conjecture of BIRCH and SWINNERTON-DYER, which connected the behaviour of the zeta-function of an elliptic curve at $s=1$ with its number-theoretic properties. Here the zeta-function is inescapably involved, both because its behaviour at $s=1$ can only be obtained by analytic continuation and because the conjecture involves not only its leading coefficient but the order of its pole at $s=1$.

Most of the subsequent work is due to, or inspired by, TATE. He has put forward, and produced evidence for, conjectures which extract a good deal of information from the zeta-function of any algebraic variety; and for surfaces over finite fields he and ARTIN have gone far towards proving these conjectures.

2. Elliptic curves over Q

We define an 'elliptic curve' to be an Abelian variety of dimension 1. Thus an elliptic curve over a field k is a complete non-singular curve of genus 1 which is defined over k and which contains a distinguished point $\mathfrak{o}$ also defined over k; the curve then has the structure of an additive group whose zero is $\mathfrak{o}$, and the group law is defined over k. We shall denote by Γ an elliptic curve over $\mathbf{Q}$, the field of rational numbers. The restriction to $\mathbf{Q}$ is needed for the computations described in §§ 3 and 4; the theoretical results quoted in the present section hold, with trivial changes of notation, over an arbitrary algebraic number field.

To state Conjectures A and B, we must first define a satisfactory global L-series, or, which comes to the same thing, a satisfactory global zeta-function associated with Γ. The crude way to produce a global zeta-function is to multiply together the local zeta-functions for all 'good' primes; this gives an Euler product with finitely many factors apparently missing. For an elliptic curve, though not for a general Abelian variety, the correct form for these missing factors is known. However, the conjectures are only concerned with the behaviour of the zeta-function near $s=1$, and instead of supplying the missing factors as functions of s it is therefore only necessary to supply their values at $s=1$. We now show how to do this.

Let ω be a differential of the first kind on Γ. If Γ is written in the traditional form

$$y^2 = x^3 - Ax - B$$

then we can choose $\omega = dx/2y$; in any case ω is unique up to multiplication by a non-zero rational number, and the choice of this number does not affect what follows. For almost all primes p, Γ and ω both have non-degenerate reductions modulo p; hence they give rise to an elliptic curve Γ_p defined over $GF(p)$, the finite field of p elements, and a differential of the first kind ω_p on Γ_p. The zeta-function of Γ_p over $GF(p)$ is

$$\zeta_p(\Gamma_p, s) = \frac{(1-\alpha_p p^{-s})(1-\bar{\alpha}_p p^{-s})}{(1-p^{-s})(1-p^{1-s})}; \tag{2.1}$$

here α_p, $\bar{\alpha}_p$ are defined by the statement that for $q=p^n$ there are just

$$N_q = p^n - \alpha_p^n - \bar{\alpha}_p^n + 1 \tag{2.2}$$

points on Γ_p defined over $GF(q)$. From the local zeta-functions (2.1) we can form a crude global L-series

$$L(\Gamma, s) = \prod \{(1 - \alpha_p p^{-s})(1 - \bar{\alpha}_p p^{-s})\}^{-1},$$

the product being taken over those primes p for which Γ has a good reduction modulo p. The product is only known to converge in $\Re s > \frac{3}{2}$; but Weil has conjectured that $L(\Gamma, s)$ can be analytically continued over the whole complex plane, and that there is a functional equation connecting $L(s)$ and $L(2-s)$. In what follows, it will be assumed that $L(s)$ is at any rate well defined in some neighbourhood of $s=1$.

Let S be a finite set of primes which contains the infinite prime and any finite prime modulo which Γ or ω does not have a good reduction. For any finite prime p, whether in S or not, we can form

$$M_p(\Gamma) = \int_{\Gamma(\mathbf{Q}_p)} |\omega|_p \mu_p \tag{2.3}$$

where $\mathbf{Q}_p$ is the field of p-adic numbers, $\Gamma(\mathbf{Q}_p)$ is the set of points on Γ defined over $\mathbf{Q}_p$, $|\ |_p$ is the usual p-adic valuation, and μ_p is the usual Haar measure on $\mathbf{Q}_p$ which assigns measure 1 to the set of p-adic integers. For the infinite prime we replace (2.3) by

$$M_\infty(\Gamma) = \int \omega$$

taken over all real points on Γ. Now write

$$L_S^*(\Gamma, s) = \prod_{p \in S} \{M_p(\Gamma)\}^{-1} \prod_{p \notin S} \{(1 - \alpha_p p^{-s})(1 - \bar{\alpha}_p p^{-s})\}^{-1}. \tag{2.4}$$

It is easily verified that for fixed S this does not depend on the particular choice of ω. Moreover, it does not depend significantly on the choice of S; for if S_1, S_2 are two such sets then

$$L_{S_1}^*(\Gamma, s)/L_{S_2}^*(\Gamma, s) \to 1 \quad \text{as} \quad s \to 1.$$

To prove this it is enough to note that if Γ and ω both have a good reduction modulo p then

$$M_p(\Gamma) = N_p/p = (1 - \alpha_p p^{-1})(1 - \bar{\alpha}_p p^{-1}). \tag{2.5}$$

Henceforth we shall denote by $L^*(\Gamma, s)$ any Euler product which differs from $L(\Gamma, s)$ only in finitely many factors and which satisfies

$$L^*(\Gamma, s)/L_S^*(\Gamma, s) \to 1 \quad \text{as} \quad s \to 1\,. \tag{2.6}$$

Note that we do not adopt the usual normalization condition that $L^*(\Gamma, s) \to 1$ as $s \to +\infty$, preferring instead to normalize at $s=1$.

We next define the Tate-Šafarevič group Ш and the Weil-Chatelet group WC; only the former of these occurs in the conjectures, but it is natural to describe both together. Let $\mathscr{S}$ be the set of all principal homogeneous spaces over Γ; in other words, an element of $\mathscr{S}$ is a complete non-singular curve C of genus 1 defined over $\mathbf{Q}$, together with a specific identification of Γ with its Jacobian. (Note that because Γ has non-trivial automorphisms, the curve C does not in itself determine the canonical map $C \times C \to \Gamma$.) For any two elements C_1, C_2 of $\mathscr{S}$, write $C_1 \sim C_2$ if there exists a birational map $C_1 \to C_2$ defined over $\mathbf{Q}$ for which the diagram

$$\begin{array}{ccc} C_1 \times C_1 & \to & C_2 \times C_2 \\ & \searrow \quad \Gamma \quad \swarrow & \end{array}$$

is commutative. This defines an equivalence relation in $\mathscr{S}$; and the set of equivalence classes in $\mathscr{S}$ is called the Weil-Chatelet set associated with Γ. It can be given the structure of a commutative group in a natural way; for the details see WEIL [*31*]. This group is a torsion group, and its identity element is the equivalence class consisting of all curves C which contain a rational point. Like most sets which unexpectedly have a natural group structure, it can also be defined as a cohomology group; it is $H^1(G, \Gamma(\overline{\mathbf{Q}}))$, where $\overline{\mathbf{Q}}$ is the algebraic closure of $\mathbf{Q}$ and G is the Galois group of $\overline{\mathbf{Q}}/\mathbf{Q}$.

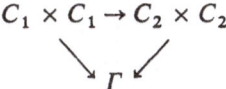

Now consider those curves C in $\mathscr{S}$ which contain points defined over each p-adic field including the reals; these are just the curves which can play a significant part in an 'infinite descent' argument applied to Γ. Such curves precisely fill a certain number of equivalence classes in $\mathscr{S}$, and these classes form a subgroup of WC called the Tate-Šafarevič group Ш. Unfortunately, very little is known about Ш, and there is no curve Γ for which it has completely determined. It is conjectured that it is always a finite group; and CASSELS [*5*] has shown that when it is finite its order must be a perfect square. It can be shown that Ш only contains finitely many elements of any given order; and the bound for the number of these elements, for a given Γ, is moderate and in principle constructive. There is no certain way of finding these elements, because there is no certain way of finding whether two curves in $\mathscr{S}$ are equivalent; in effect, this comes down to determining whether a given curve C has a rational point. However, in practice one can usually find the elements of order 2 in Ш for any particular elliptic curve Γ, and the elements of order 3 or 4 for suitably chosen curves Γ, without intolerable labour.

In the description which follows, it will be implicitly assumed that Ш is finite. Conjectures A, B and C below, as they are stated here, do indeed each imply that Ш is finite; for each of them contains a formula in which the order of Ш appears. However, the evidence for these conjectures cannot really be regarded as evidence for the finiteness of Ш. In fact, each of these formulae contains a term which is certainly associated with Ш and whose value appears to be always a positive integer – and indeed a perfect square. It is natural to identify this term with the order of Ш; but one can give a more complicated interpretation of it which is equally compatible with the evidence and which would still make sense even if Ш were sometimes infinite.

MORDELL [*16*] has shown that $\Gamma(\mathbf{Q})$, the group of rational points on Γ, is finitely generated; and in any particular case his proof gives an explicit bound for the number of generators. It is easy to find the elements of finite order in $\Gamma(\mathbf{Q})$; see for example CASSELS [*6*], Theorem 22.1. There is no certain method of calculating g, the rank of $\Gamma(\mathbf{Q})$; but in practice one can usually find g without intolerable labour by the method of infinite descent. BIRCH and SWINNERTON-DYER [2] have given an alternative method of conducting the first descent, which avoids any use of algebraic number fields and is therefore suitable for machine computation. Moreover, on the assumption that Ш is finite CASSELS [*5*] has shown that the difference between the number of first descents and the value of g is even; thus the parity of g can be found even when g itself cannot.

We can now state the initial conjecture of BIRCH and SWINNERTON-DYER [*3*]. This can be regarded primarily as giving a necessary and sufficient condition for $g=0$ – that is, for $\Gamma(\mathbf{Q})$ to be finite.

Conjecture A. *With the notation and definitions above,*

$$L^*(\Gamma, 1) = \begin{cases} [\text{Ш}]/[\Gamma(\mathbf{Q})]^2 & \textit{if } g=0, \\ 0 & \textit{otherwise}, \end{cases}$$

where square brackets denote the order of a group.

To produce a more detailed conjecture when $g>0$ we need a measure of the density of the rational points on Γ. We first define the height of a rational point on Γ, following the original approach of TATE as described in [*15*]. Assume that Γ is embedded in projective space, and fix a system of co-ordinates in that space. For any rational point P on Γ, write

$$h(P) = \log(\max\{|x_0|, |x_1|, \ldots, |x_n|\})$$

where $(x_0, x_1, \ldots, x_n)$ is that representation of P for which the x_i are integers with no common factor. If nP denotes the sum of n copies of P under the standard addition law on Γ, then

$$\hat{h}(P) = \lim_{n\to\infty} n^{-2} h(nP)$$

exists and is called the height of P. Moreover $\hat{h}(P)$ behaves like a quadratic form, does not depend on the choice of a co-ordinate system in the ambient projective space, and vanishes at just those points of $\Gamma(\mathbf{Q})$ which are of finite order. We can derive from $\hat{h}(P)$ the bilinear form

$$\hat{h}(P, P') = \tfrac{1}{2}\{\hat{h}(P+P') - \hat{h}(P) - \hat{h}(P')\}.$$

The effect of a birational transformation of Γ is to multiply $\hat{h}$ by a constant; henceforth we normalize $\hat{h}$ by taking Γ to be a non-singular plane cubic curve. Birch has given an explicit formula (quoted in [*25*]) for $\hat{h}(P)$ for curves of the form

$$x^3 + y^3 = Dz^3 \tag{2.7}$$

and presumably this can be modified to be valid for all elliptic curves.

Now let $P_1, \ldots, P_g$ be a base for $\Gamma(\mathbf{Q})$ modulo torsion, and write

$$R = \det\{\hat{h}(P_i, P_j)\};$$

it is easy to show that R is positive and does not depend on the choice of base. The natural generalization of Conjecture A is as follows; it was first explicitly stated in Stephens [*24*], though its general shape had been suggested earlier.

Conjecture B. *With the notation and definitions above,*

$$L^*(\Gamma, s) \sim \frac{[\text{Ш}]\, R}{[\Gamma(\mathbf{Q})_{\text{tors}}]^2}(s-1)^g \quad as \quad s \to 1.$$

This of course includes Conjecture A; but so much of the evidence applies to the special case of Conjecture A that it is convenient to give both statements explicitly. There is no curve Γ for which Conjecture B has been proved to hold, because there is no curve for which Ш is completely determined; however, the supporting evidence is very strong. The direct evidence is of three kinds:

(i) invariance under isogeny;

(ii) calculations of $L^*(\Gamma, s)$ near $s=1$;

(iii) deductions from the functional equation of $L^*(\Gamma, s)$. Moreover, there is some analogy with the work of SIEGEL and TAMAGAWA on quadratic forms, and a close analogy with the work of ARTIN and TATE on surfaces over finite fields described in § 6 below.

Let Γ' be an elliptic curve which is isogenous to Γ over Q. Cassels has shown that if Conjecture B holds for Γ it also holds for Γ'. All the terms in Conjecture B except for g are liable to change under isogeny. However, in each case the ratio between the values of a term for Γ and for Γ' is easier to determine than the two values themselves; in particular, the change in $L^*(\Gamma, s)$ comes entirely from the factors $M_p(\Gamma)$ corresponding to the bad primes. CASSELS' result is therefore a purely algebraic one, which does not involve s and which in particular throws no light on the problem of analytic continuation.

Experiment shows that $L^*(\Gamma, 1)$ cannot be satisfactorily calculated either by numerical analytic continuation or by setting $s=1$ in (2.4) and truncating the infinite product at a suitable point. It is known however that $L^*(\Gamma, s)$ can be analytically continued across $\Re s = \frac{3}{2}$ when one of two conditions holds; and in these cases $L^*(\Gamma, 1)$ can be calculated. These conditions are that Γ admits complex multiplication, discussed in § 3, or that Γ can be parametrized by elliptic modular functions, discussed in § 4. Here we consider just what the resulting figures show. Write

$$L^*(\Gamma, s) = a_0 + a_1(s-1) + a_2(s-1)^2 + \cdots .$$

It can be shown, under either condition, that $a_0 = L^*(\Gamma, 1)$ is a rational number, for whose denominator an explicit bound can be given; since it can be computed to any desired accuracy, it can be found exactly even though the computations themselves are not exact. By contrast, it is only practicable to compute the a_n with $n>0$ to two or three significant figures; and no theoretical statement about them is known. For curves with complex multiplication the numerical results are given in [*3*] and [*24*]. Of the 2024 curves for which $L^*(\Gamma, 1)$ has been calculated, there are 1744 for which the value of g is also known. Of these, 984 have $L^*(\Gamma, 1)=0$ and $g>0$, in accordance with Conjecture A; the other 760 have $g=0$ and $L^*(\Gamma, 1)>0$, and for these it is natural to use the formula of Conjecture A to give a hypothetical value of $[Ш]$. In each case the value is a perfect square, in accordance with the result of CASSELS [*5*]. Moreover, for each such curve either the 2-component or the 3-component of Ш was found in the course of finding g; and these agree with the hypothetical values of $[Ш]$. There are 355 curves for which $g>0$ and both R and the a_n with $n>0$ have been calculated. These support the stronger Conjecture B, with the same interpretation of $[Ш]$ as above, and with two minor reservations. Since the values of a_g and R are only approximate, so is that which is obtained for $[Ш]$ from the formula. However this value is always close to a small integer, and so the hypothetical value of $[Ш]$ is unambiguous. Again, for $g>1$ Conjecture B requires that $a_1=0$ and direct calculation can only show that this holds approximately. When $g=2$ we can use the functional equation for $L^*(\Gamma, s)$ and the exactly verifiable statement $a_0=0$ to prove that $a_1=0$; but in the two relevant cases when $g=3$ there is no known way to prove $a_1=0$, though the calculations make this plausible. As yet, there is only one curve without complex multiplication for which the value of $L^*(\Gamma, 1)$ is known. This case is fully described in § 4; it supports Conjecture A provided that Ш is trivial, which there is no reason to doubt.

In describing the functional equation, it is convenient to write

$$\Lambda^*(s) = (2\pi)^{-s}\,\Gamma(s)\,L^*(\Gamma, s). \tag{2.8}$$

WEIL has conjectured that for a suitable choice of the factors corresponding to the bad primes,

$$\Lambda^*(s) = \varepsilon f^{1-s} \Lambda^*(2-s). \tag{2.9}$$

Here $\varepsilon=\pm 1$ and f is the conductor of Γ, so that f is a product of suitable powers of the finite bad primes; for the precise definition of f see [*17*] or [*19*]. In general, no formula for ε is known; but DEURING has shown that if Γ admits complex multiplication then $L^*(\Gamma, s)$ is essentially a Hecke L-series with Großencharaktere, and hence (2.9) holds with an explicitly defined ε. Evidently $L^*(\Gamma, s)$ has a zero of odd order at $s=1$ if $\varepsilon=-1$, and of even order if $\varepsilon=+1$. According to Conjecture B, $L^*(\Gamma, s)$ has a zero of order g at $s=1$. For Γ of the form

$$y^2 = x^3 - Dx \tag{2.10}$$

or of the form (2.7), which correspond to the two simplest cases of complex multiplication, BIRCH and STEPHENS [*1*] have shown that the number of first descents is even when $\varepsilon=+1$ and odd when $\varepsilon=-1$; this supports the conjecture in view of the theorem of CASSELS quoted above, that provided Ш is finite the number of first descents has the same parity as g.

NERON [*17*] has given a birationally invariant theory of the reduction of elliptic curves modulo p. This appears, inter alia, to give the right form for the factors of $L^*(\Gamma, s)$ corresponding to the bad primes; and I am indebted to TATE for pointing out to me that it also gives a simpler definition of the corresponding factors $M_p(\Gamma)$ than that of (2.3). In fact NERON shows that there is an essentially unique model for Γ of the form

$$Y^2 + \lambda XY + \mu Y = X^3 + \alpha X^2 + \beta X + \gamma \tag{2.11}$$

where $\lambda, \mu, \alpha, \beta, \gamma$ are integers and the discriminant of the equation (2.11) is as small as possible. For this model, every Γ_p is an irreducible curve. Write

$$L_p(s) = \{(1 - \alpha_p p^{-s})(1 - \bar{\alpha}_p p^{-s})\}^{-1}$$

with the $\alpha_p, \bar{\alpha}_p$ of (2.1) if Γ_p is non-singular;

$$L_p(s) = (1 - p^{-s})^{-1}$$

if Γ_p has an ordinary double-point with distinct rational tangents;

$$L_p(s) = (1 + p^{-s})^{-1}$$

if Γ_p has an ordinary double point with irrational tangents; and

$$L_p(s) = 1$$

if Γ_p has a cusp. Moreover, let c_p be the number of irreducible components of multiplicity 1 in the Neron fibre associated with the reduction of (2.11) modulo p; thus $c_p = 1$ whenever Γ_p is non-singular. TATE has shown that

$$M_p(\Gamma) = c_p / L_p(1)$$

for all primes p, which generalizes (2.5). In view of this, the correct definition of $L^*(\Gamma, s)$ is presumably

$$L^*(\Gamma, s) = \{M_\infty(\Gamma) \prod c_p\}^{-1} \prod L_p(s).$$

Here the term in curly brackets has been introduced to preserve (2.6). According to SERRE, this is the L-series for which the functional equation (2.9) should hold.

3. Elliptic curves with complex multiplication

In this section we give an account of the calculation of $L^*(\Gamma, s)$ for curves of the form

$$y^2 = x^3 - Dx \tag{3.1}$$

where D is a rational integer which we can take to be fourth-power-free. Similar arguments apply to the other types of curve defined over $\mathbf{Q}$ which admit complex multiplication, but the formulae differ. Detailed calculations have also been carried out by STEPHENS [24] for the curves (2.7), and pupils of CASSELS are currently working on curves of the form

$$y^2 = x^3 + 4Ax^2 + 2A^2x$$

which admit complex multiplication by $\sqrt{(-2)}$.

For the curve (3.1) the bad primes are just those which divide $2D$. For any other prime p it is known that

$$N_p = \begin{cases} p + 1 - \bar{\pi}\left(\frac{D}{\pi}\right)_4 - \pi\left(\frac{D}{\bar{\pi}}\right)_4 & \text{for} \quad p \equiv 1 \bmod 4, \\ p + 1 & \text{for} \quad p \equiv 3 \bmod 4, \end{cases}$$

where in the upper line $(\ )_4$ denotes the biquadratic residue symbol in $\mathbf{Q}(i)$ and $\pi, \bar{\pi}$ are primes in $\mathbf{Q}(i)$ such that $p = \pi\bar{\pi}$ and

$$\pi \equiv \bar{\pi} \equiv 1 \quad \bmod (2 + 2i).$$

It follows after a little manipulation that

$$L(\Gamma, s) = \prod \left\{1 - \left(\frac{D}{\pi}\right)_4 \frac{\bar{\pi}}{(N\pi)^s}\right\}^{-1} = \sum \left(\frac{D}{\sigma}\right)_4 \frac{\bar{\sigma}}{(N\sigma)^s}, \tag{3.2}$$

where N denotes the norm from $\mathbf{Q}(i)$ to $\mathbf{Q}$, the product is taken over all real or complex Gaussian primes and the sum over all Gaussian integers subject to

$$\pi \equiv \sigma \equiv 1 \quad \bmod (2 + 2i)$$

in each case. Now let $\Delta \equiv 1 \bmod (2+2i)$ be the odd part of the square-free kernel of D, and write in (3.2)

$$\sigma = 16\Delta\mu + \varrho$$

where μ runs through all Gaussian integers and ϱ runs through a certain finite set. This gives, writing $\alpha = \varrho/16\Delta$ for convenience,

$$L(\Gamma, s) = (16\Delta)^{1-2s} \sum \sum \left(\frac{D}{\varrho}\right)_4 \frac{\bar{\mu} + \bar{\alpha}}{|\mu + \alpha|^{2s}}. \tag{3.3}$$

The sum over μ does not involve the biquadratic residue symbol, and it can be analytically continued into $\Re s > \frac{1}{2}$ as follows. Write

$$\psi(\alpha, s) = \frac{\bar{\alpha}}{|\alpha|^{2s}} + \sum_{\mu \neq 0} \left\{\frac{\bar{\alpha} + \bar{\mu}}{|\alpha + \mu|^{2s}} - \frac{\bar{\mu}}{|\mu|^{2s}}\left(1 - \frac{s\alpha}{\mu} + \frac{\bar{\alpha}(1-s)}{\bar{\mu}}\right)\right\}. \tag{3.4}$$

Since the expression in curly brackets is $O(\mu^{-2s-1})$ this defines an analytic function in $\Re s > \frac{1}{2}$; and moreover

$$\psi(\alpha, 1) = \frac{1}{\alpha} + \sum_{\mu \neq 0} \left\{ \frac{1}{\mu + \alpha} - \frac{1}{\mu} + \frac{\alpha}{\mu^2} \right\}$$

is just the Weierstrass zeta-function with periods 1, i. On the other hand, if $\Re s > \frac{3}{2}$ we can rearrange (3.4) to give

$$\sum_{\mu} \frac{\bar{\alpha} + \bar{\mu}}{|\alpha + \mu|^{2s}} = \psi(\alpha, s) + \bar{\alpha}(1 - s) \sum_{\mu \neq 0} \frac{1}{(N\mu)^s} =$$
$$= \psi(\alpha, s) + 4\bar{\alpha}(1 - s)\, \zeta_{\mathbf{Q}(i)}(s),$$

the other terms in the sum cancelling in pairs. Substituting into (3.3) we obtain

$$L(\Gamma, s) = (16\Delta)^{1-2s} \sum_{\varrho} \left\{ \left(\frac{D}{\varrho}\right)_4 \psi\left(\frac{\varrho}{16\Delta}, s\right) + \frac{(1 - s)\, \zeta_{\mathbf{Q}(i)}(s)}{4\Delta}\, \bar{\varrho} \left(\frac{D}{\varrho}\right)_4 \right\}. \tag{3.5}$$

There are now two ways to proceed. For Conjecture A, we are only interested in the value of $L(\Gamma, 1)$ and can simply write $s = 1$ in (3.5). This gives a closed expression for $L(\Gamma, 1)$ in terms of values of the Weierstrass elliptic functions with periods 1, i. The resulting expression for $L^*(\Gamma, 1)$ is well suited to computation; moreover since the number-theoretic properties of division values of the Weierstrass $\wp$-functions are well known, one can show that $L^*(\Gamma, 1)$ is rational and can give an explicit bound for its denominator. For the full details, see [3].

For Conjecture B, on the other hand, we need to find the first few coefficients in the power series expansion of $L(\Gamma, s)$ about $s = 1$. It turns out that except when $\Delta = 1$, for which special devices are needed, we can arrange that

$$\sum \left(\frac{D}{\varrho}\right)_4 = \sum \varrho \left(\frac{D}{\varrho}\right)_4 = \sum \bar{\varrho} \left(\frac{D}{\varrho}\right)_4 = 0.$$

Assuming this, we can argue back from (3.5) and (3.4) in a way which shows that (3.3) converges in $\Re s > \frac{1}{2}$ provided that the inner sum is taken over ϱ and the outer sum over μ. From this we can deduce convergent series for the coefficients of the power series expansion of $L(\Gamma, s)$ about $s = 1$. Unfortunately the convergence is not very strong; even if one uses devices from numerical analysis to accelerate the convergence, a large amount of computer time is needed and the results are only accurate to two or three significant figures. These calculations have not in fact been carried out for curves of type (3.1), because the computer available when the work reported in [3] was done was not fast enough. However, STEPHENS has carried out the corresponding calculations for curves of the form (2.7), on the Atlas I at Manchester University.

Curves of the form (3.1) can be parametrized by modular functions, and the methods described in the next section can therefore be applied to them. These methods have only recently become available, and one has too little experience of them yet to know how much labour they involve. At the moment it seems that the methods of the present section are preferable for calculating $L(\Gamma, 1)$, but those of the next section are preferable if further coefficients in the power series expansion of $L(\Gamma, s)$ are needed.

There is a third method of computing $L(\Gamma, s)$ which deserves mention, though as yet no one has tried to exploit it. We have already pointed out that for curves with complex multiplication $L(\Gamma, s)$ is a Hecke L-series with Großencharaktere. HECKE [12] has shown that every such function can be analytically continued over the whole plane, by expressing it in terms of integrals involving theta-functions. These integrals are quite convenient for numerical calculation. This approach has the disadvantage that it does not yield a closed formula for $L(\Gamma, 1)$, and hence all the results it produces are approximate. But it has the advantage that in principle it can be carried through for curves Γ defined over an arbitrary algebraic number field, provided they admit complex multiplication. In contrast, the methods of the present section applied to (3.1) work only if the ground field is $\mathbf{Q}$ or $\mathbf{Q}(i)$, and the methods of the next section apply only to curves defined over $\mathbf{Q}$.

4. Elliptic curves parametrized by modular functions

The work reported in this section has only been started very recently; consequently the calculations are incomplete and most of the proofs are missing. It is based on a conjecture of WEIL [32] and on theorems of EICHLER [7] and SHIMURA [23]. There is as yet only one curve Γ for which the value of $L^*(\Gamma, 1)$ has been calculated by these methods, and the first part of this section outlines this calculation. The second part describes the general method, and those results which have so far been obtained.

We shall use in this section the usual conventions of modular function theory; in particular τ is a complex variable, H is the upper half-plane $\operatorname{Im}\tau>0$, and if $q>0$ is an integer then $\Gamma_0(q)$ is the group of transformations

$$\tau \to \frac{a\tau+b}{c\tau+d}$$

where a, b, c, d are rational integers with $ad-bc=1$ and $q|c$.

Let $j(\tau)$ be the fundamental elliptic modular function. The quotient space $H/\Gamma_0(11)$ has genus 1, and the curve

$$\Gamma: y^2 + y = x^3 - x^2 - 10x - 20 \tag{4.1}$$

is a model for the associated function field $\mathbf{C}(j(\tau), j(11\tau))$. This is the reduced model in the sense of NÉRON, and may be deduced from equation (13) of FRICKE [8], p. 406. The only bad prime for Γ is $p=11$, for which in the notation of § 2 we have

$$L_{11}(s) = (1+11^{-s})^{-1}, \qquad c_{11} = 5.$$

With the canonical differential

$$\omega = \frac{dx}{2y+1}$$

we can therefore write $L^*(\Gamma, s) = L(\Gamma, s)/5\,M_\infty(\Gamma)$ where

$$L(\Gamma, s) = (1+11^{-s})^{-1} \prod \{(1-\alpha_p p^{-s})(1-\bar{\alpha}_p p^{-s})\}^{-1} = \sum a_n n^{-s}$$

say. Now SHIMURA has shown that if $f(\tau) = \sum a_n e^{2\pi i n\tau}$ then $f(\tau)\,d\tau$ is a differential of the first kind on $H/\Gamma_0(11)$ and is therefore a multiple of ω; comparing the coefficients of $e^{2\pi i\tau}$ we obtain with the help of FRICKE

$$\omega = -2\pi i f(\tau)\,d\tau.$$

By the Mellin transform formula we therefore have

$$L(\Gamma, 1) = -2\pi i \int_0^{i\infty} f(\tau)\,d\tau = \int_{\tau=0}^{i\infty} \omega. \tag{4.2}$$

But the points $\tau=0$ and $\tau=i\infty$ correspond respectively to the points $(-6,5)$ and $(5,5)$ on Γ, both of which are 5-division points; and it now follows easily that

$$L(\Gamma, 1) = \tfrac{1}{5}\int \omega = \tfrac{1}{5} M_\infty(\Gamma), \quad L^*(\Gamma, 1) = \tfrac{1}{25},$$

in agreement with the known results $g=0$, $[\Gamma(\mathbf{Q})]=5$ and the conjectured results that Ш is trivial and that Conjecture A holds for the curve (4.1). It is no coincidence that the integral (4.2) can be explicitly evaluated in this way. On the one hand, using standard notation

$$\Delta(11\omega_1, \omega_2)/\Delta(\omega_1, \omega_2)$$

is a modular function invariant under $\Gamma_0(11)$ which has a ten-fold zero at $\tau=i\infty$ and a ten-fold pole at $\tau=0$; hence $(i\infty)-(0)$ is a 10-division point on Γ. On the other hand, x and y are rational functions of $j(\tau)$ and $j(11\tau)$ over $\mathbf{Q}$; hence their values must be rational (or infinite) at $\tau=0$ and $\tau=i\infty$, since these are both points about which $j(\tau)$ and $j(11\tau)$ have power series expansions with rational coefficients. These arguments, and the known fact that $[\Gamma(\mathbf{Q})]=5$, are enough to show that $L(\Gamma, 1)=\frac{1}{5}nM_\infty(\Gamma)$ for some integer n; and by finding the actual points on Γ which correspond to $\tau=0$ and $\tau=i\infty$ we see moreover that $n\equiv 1 \bmod 5$. The correct value of n can now be found by crude numerical estimation, or more elegantly by topological arguments based on a knowledge of where the real points of Γ correspond to on the Riemann surface $H/\Gamma_0(11)$.

There are just 12 values of q for which $H/\Gamma_0(q)$ has genus 1. The values of q and the corresponding curves Γ, in unreduced form, may be found in [*8*]. Presumably similar arguments will work for each of them.

The importance of this method, however, arises from a conjecture of WEIL. The justification of this conjecture is given in [*32*] and need not be repeated here; but it should be emphasized that the theoretical reasons for it are much more powerful than the numerical evidence reported below. Let Γ be an elliptic curve defined over $\mathbf{Q}$, and let f be its conductor; then Weil's conjecture states that Γ can be parametrized by elliptic modular functions invariant under $\Gamma_0(f)$.

Because the difficulty of describing $H/\Gamma_0(f)$ explicitly increases rapidly with f, it is convenient to start from the other end; that is, to choose $q>0$ and ask what curves Γ of genus 1 are parametrized by modular functions invariant under $\Gamma_0(q)$. If $C_0(q)$ is the curve, defined over $\mathbf{Q}$, whose Riemann surface is $H/\Gamma_0(q)$, this is equivalent to looking for maps $C_0(q)\to\Gamma$. Let g be the genus of $C_0(q)$ – the previous meaning of g will not be needed in this section – and for any differential of the first kind Ω on $C_0(q)$ and any homology class α write

$$\langle\Omega, \alpha\rangle = \int_\alpha \Omega. \tag{4.3}$$

By a slight abuse of language this can be viewed as a bilinear form, Ω being in a vector space over the complex numbers and α in a vector space over $\mathbf{Q}$; and these spaces have dimension g and $2g$ respectively. Now suppose that there is a map $C_0(q) \to \Gamma$, and let ω be the unique differential of the first kind on Γ and Ω the induced differential of the first kind on $C_0(q)$; then for any homology class α on $C_0(q)$ whose image on Γ is trivial we have

$$\langle \Omega, \alpha \rangle = \int_{\alpha} \Omega = \int_{0} \omega = 0 .$$

Such homology classes form a subspace of dimension $2g-2$. Conversely, suppose that on $C_0(q)$ there is a differential of the first kind Ω such that the α for which $\langle \Omega, \alpha \rangle = 0$ form a subspace of dimension $2g-2$. Choose a base $\alpha_1, \ldots, \alpha_{2g}$ for the integral homology of $C_0(q)$ such that $\langle \Omega, \alpha_n \rangle = 0$ for $n > 2$; then the function ϕ on $C_0(q)$ defined by

$$\phi(P) = \int^{P} \Omega$$

is many-valued, and if one of its values is $\phi_0(P)$ the others are the

$$\phi_0(P) + n_1 \langle \Omega, \alpha_1 \rangle + n_2 \langle \Omega, \alpha_2 \rangle$$

where n_1 and n_2 are arbitrary integers. If Γ is the elliptic curve corresponding to the doubly periodic functions with periods $\langle \Omega, \alpha_1 \rangle$ and $\langle \Omega, \alpha_2 \rangle$ it follows that ϕ induces a map $C_0(q) \to \Gamma$.

Hence to find the curves Γ it is enough to find differentials Ω satisfying the conditions above. Unfortunately, direct methods are no use, for it is inconvenient to form the space of differentials of the first kind and impossible to evaluate the integrals $\langle \Omega, \alpha \rangle$ exactly. To progress, we introduce the Hecke operator. For a full account of this, see HECKE [*13*] and PETERSSON [*21*]; here I quote only the results which are needed. Let W denote the space of differentials of the first kind on $C_0(q)$, and let $\Omega = f(\tau)\, d\tau$ be an element of W, so that $f(\tau)$ is a cusp form of dimension -2. For each prime p not dividing q, we define an endomorphism T_p on W by the formula

$$T_p \Omega = \left\{ p f(p\tau) + p^{-1} \sum_{n=0}^{p-1} f\left(\frac{\tau + n}{p}\right) \right\} d\tau .$$

(There are similar operators T_p^* for the p which divide q; they play a part in the complete theory, but are omitted here for simplicity.) The T_p commute, and we can choose a base for W each of whose members is an eigenvector for each T_p. In view of the importance of the bilinear form (4.3), we ought to define a dual operator T_p' on V, the space of homology classes α on $C_0(q)$. Define T_p' as the map induced on V by the map of 0-cycles

$$(\tau) \to (p\tau) + \sum\left(\frac{\tau+n}{p}\right);$$

then it is easily verified that the T_p' are well defined, commute, and satisfy

$$\langle T_p\Omega, \alpha\rangle = \langle \Omega, T_p'\alpha\rangle. \tag{4.4}$$

Again, the map $\tau \to -\bar{\tau}$ induces a homeomorphism of $H/\Gamma_0(q)$ viewed merely as a topological space, and hence induces an automorphism of V which evidently commutes with each T_p. Let V^+ be the subspace consisting of those α which are fixed under this automorphism, and V^- the subspace of those α which are reversed in sign; then $V = V^+ \oplus V^-$, the T_p' induce endomorphisms of V^+ and V^-, and $\mathbf{C} \otimes V^+$ and $\mathbf{C} \otimes V^-$ are canonically dual to W, where $\mathbf{C}$ denotes the field of complex numbers. In particular, the eigenvalues of T_p for W, and of T_p' for V^+ and V^- are the same. Now let α^+ be an isolated eigenvector for the T_p' acting on V^+ – that is, an eigenvector which is determined up to multiplication by a constant by its eigenvalues. It is easy to see that there are corresponding eigenvectors Ω in W and α^- in V^-, and by (4.4) that the α in V for which $\langle \Omega, \alpha\rangle = 0$ form a subspace of dimension $2g-2$; hence α^+ induces a map $C_0(q) \to \Gamma$. Conversely, if we are given a map $C_0(q) \to \Gamma$ it can be shown that the corresponding Ω is an eigenvector for every T_p and that the corresponding eigenvalues are rational integers. Non-isolated eigenvectors do occur, but apparently they correspond to the proper factors q' of q, for each of which there exist several distinct canonical maps $C_0(q) \to C_0(q')$; hence they are not important.

It is a straightforward matter to find a base for V^+, say, and to compute the matrix which represents the effect of any T_p' on it; and in this way we can easily find, for any given q, the α^+ which induce maps $C_0(q) \to \Gamma$. Let

$$T_p'\alpha^+ = c_p\alpha^+$$

for each p, and let the corresponding differential be

$$\Omega = f(\tau)\,d\tau \quad \text{where} \quad f(\tau) = \sum a_n e^{2\pi i n\tau}.$$

Hecke [13] has shown that

$$\sum a_n n^{-s} = F \prod (1 - c_p p^{-s} + p^{1-2s})^{-1} \tag{4.5}$$

where F is a factor corresponding to the primes which divide q. In principle we can now find the two non-zero periods of Ω by numerical integration, and hence also the curve Γ; however, this method is unattractive and there is no guarantee that the resulting curve Γ will have integral coefficients. We have preferred to look for elliptic curves of conductor q and pair them off empirically with the isolated eigenvectors α^+. Despite some special results of OGG [20] and others, there is no certain way of finding all curves of conductor q. Instead, I have written a search program which examines all curves

$$y^2 + b_1 xy + b_3 y = x^3 + b_2 x^2 + b_4 x + b_6$$

with $b_1 = 0$ or 1, $b_2 = 0$ or ± 1, $b_3 = 0$ or 1, $|b_4| < 300$ and $|b_6| < 1000$. For each $q \leqslant 75$ the search program produces exactly as many non-isogenous curves of conductor q as there are isolated rational eigenvectors α^+. Moreover, it is easy to pair them off; for although no proof is yet available there are strong theoretical and numerical reasons for supposing that (4.5) is the L-series associated with Γ.

Assuming all this, the Mellin transform theorem gives

$$L(\Gamma, 1) = \int_0^{i\infty} -2\pi i \Omega,$$

which is equal to the integral of a multiple of ω, the unique differential of the first kind on Γ, along a certain contour on Γ. As in the special case worked out at the beginning of this section, the ends of this contour will be rational points on Γ, and they will differ by a $(q-1)$-division point. Hence $L^*(\Gamma, 1)$ will be a rational number for whose denominator an explicit bound can be given. Similarly, the coefficients in the power series expansion of $L^*(\Gamma, s)$ about $s=1$ can be obtained as definite integrals suitable for numerical calculation. However, the detailed programming of these calculations is a slow job.

5. Abelian varieties over an algebraic number field

It is natural to try to generalize Conjectures A and B, both by replacing the elliptic curve Γ by an Abelian variety A and by replacing the ground field Q by an arbitrary algebraic number field κ. This has been done by TATE [29], and the result is Conjecture C below. It is known that the underlying theory remains much the same, though the proofs of the main theorems become harder.

Let A be an Abelian variety of dimension d, defined over an algebraic number field κ. For almost all primes $\mathfrak{p}$ of κ, A has a non-degenerate reduction $A_\mathfrak{p}$ modulo $\mathfrak{p}$ which is an Abelian variety defined over the finite field $GF(q)$ of $q = N\mathfrak{p}$ elements. Moreover there exist numbers $\alpha_{1\mathfrak{p}}, \ldots, \alpha_{2d\mathfrak{p}}$, each of absolute value $q^{1/2}$, such that the number of points on $A_\mathfrak{p}$ defined over $GF(q^n)$ is $\prod(1-\alpha_{i\mathfrak{p}}^n)$ for each n. From these we define the local L-series

$$L_\mathfrak{p}(s) = \{\prod(1-\alpha_{i\mathfrak{p}}q^{-s})\}^{-1}.$$

Now let ω be a non-zero invariant exterior differential form of degree d on A, let $\kappa_\mathfrak{p}$ be the field of $\mathfrak{p}$-adic numbers, $A(\kappa_\mathfrak{p})$ the set of points on A defined over $\kappa_\mathfrak{p}$, $|\ |_\mathfrak{p}$ the usual $\mathfrak{p}$-adic valuation and $\mu_\mathfrak{p}$ the usual Haar measure on $\kappa_\mathfrak{p}$ which assigns measure 1 to the $\mathfrak{p}$-adic integers. For any finite prime $\mathfrak{p}$ we write

$$M_\mathfrak{p}(A) = \int_{A(\kappa_\mathfrak{p})} |\omega|_\mathfrak{p}\mu_\mathfrak{p}^d;$$

and we make a similar definition for the infinite primes. If ω and A both have good reductions modulo $\mathfrak{p}$ then

$$M_\mathfrak{p}(A) = \{L_\mathfrak{p}(1)\}^{-1}.$$

To define a global L-series for A, choose any finite set of primes S which includes all the infinite primes and all primes for which ω or A has a bad reduction; and write

$$L_S^*(A,s) = \prod_{\mathfrak{p}\in S}\{M_\mathfrak{p}(A)\}^{-1}\prod_{\mathfrak{p}\notin S}L_\mathfrak{p}(s).$$

This depends on S, but not in any vital way. Presumably there is a best possible form for the L-series, as there is with elliptic curves, but the details are not known. $L^*(A,s)$ is only defined in $\Re s > \frac{3}{2}$, but it is conjectured that it can be analytically continued over the whole complex plane.

The Tate-Šafarevič group Ш is defined as in § 2. It is conjectured to be finite, and TATE [26] has shown that if it is finite its order is a perfect square. WEIL [30] proved that $A(\kappa)$, the group of points on A defined over κ, is finitely generated; denote by g the rank of this group. Now denote by $\hat{A}$ the Abelian variety dual to A; $\hat{A}$ is isogenous to A and so the groups $A(\kappa)$ and $\hat{A}(\kappa)$ must have the same rank, but they need not have the same torsion part. The canonical height is now defined as a bilinear function

$$\hat{h}: \hat{A}(\kappa) \times A(\kappa) \to \text{reals}.$$

It can be defined by methods similar to those of § 2, but it is preferable to use the ideas of NERON [*18*]. Now let $P_1, \ldots, P_g$ be a base for $A(\kappa)$ modulo torsion, and $\hat{P}_1, \ldots, \hat{P}_g$ be a base for $\hat{A}(\kappa)$ modulo torsion, and write

$$R = \det(\hat{h}(P_i, \hat{P}_j)).$$

Moreover let D be the discriminant of κ, and r the number of its complex infinite primes. TATE's generalization of Conjecture B is as follows:

Conjecture C. *With the notation and definitions above,*

$$L^*(A, s) \sim \frac{(2^r |D|^{-\frac{1}{2}})^d [\text{Ш}] |R|}{[A(\kappa)_{\text{tors}}] [\hat{A}(\kappa)_{\text{tors}}]} (s-1)^g \quad \textit{as} \quad s \to 1.$$

The main surprise in this, in comparison with Conjecture B, is the appearance of the dual Abelian variety $\hat{A}$ in the denominator; for Γ, as a Jacobian, is canonically self-dual. The justification for it has been provided by TATE [*29*], who showed that Conjecture C is compatible with isogeny in its present form and would not be if $\hat{A}$ was replaced by A.

There is no direct evidence for Conjecture C beyond that which applies to the special case of Conjecture B, for no way of calculating $L^*(A, s)$ near $s=1$ is known. However, the method suggested at the end of § 3 could in principle be applied to varieties with sufficiently many complex multiplications, in view of the results of SHIMURA and TAMIYAMA [*22*].

For later reference, it is convenient to rephrase the special case in which A is the Jacobian of a curve C which is also defined over κ. The L-series can be defined in terms of the local zeta-functions

$$\{\prod(1 - \alpha_{ip} q^{-s})\}/(1 - q^{-s})(1 - q^{1-s})$$

of C, and $A(\kappa) = \hat{A}(\kappa)$ is just the group of divisors on C of degree 0 and defined over κ, modulo linear equivalence. The definition of Ш in § 2 becomes meaningless, but the definition by means of cohomology groups is easily extended; and similar remarks apply to $\hat{h}$ and hence to R. All the expressions in the Conjecture can therefore be expressed in terms of the curve C.

6. Varieties over finite fields

Let V be a complete non-singular variety of dimension d, defined over the finite field $k=GF(q)$ of characteristic p. The zeta-function of V can be written in the form

$$\zeta_V(s) = \frac{P_1(q^{-s}) \dots P_{2d-1}(q^{-s})}{P_0(q^{-s})\, P_2(q^{-s}) \dots P_{2d}(q^{-s})} \tag{6.1}$$

where the P_i are polynomials (possibly of degree zero)

$$P_i(x) = \prod_{j=1}^{B_i} (1 - \alpha_{ij} x). \tag{6.2}$$

The original definition of the α_{ij} was that for each $n>0$ the number of points on V defined over $GF(q^n)$ is $\sum\sum(-1)^i \alpha_{ij}^n$; and they are assigned to the polynomials $P_i(x)$ by the conjectural relation

$$|\alpha_{ij}| = q^{i/2}. \tag{6.3}$$

Alternatively, one can define $P_i(x)$ as the characteristic polynomial of the Frobenius endomorphism acting on the i-dimensional cohomology of V with l-adic integer coefficients, where l is any prime other than p. It is believed that the $P_i(x)$ defined in this way do not depend on l; but this has only been proved for $i=0$, 1, $2d-1$ and $2d$, and of course for $\zeta_V(s)$ as defined by (6.1). In what follows we shall assume that the $P_i(x)$ are well-defined, but we shall not need (6.3).

Let ϱ_r be the rank of the group of classes of r-dimensional cycles defined over k on V, modulo algebraic equivalence. We can choose a base for the $2r$-dimensional cohomology of V such that ϱ_r of its elements come from these cycles. Since the Frobenius endomorphism acts trivially on these cycles, the characteristic polynomial $P_{2r}(x)$ must contain the factor $(1-q^r x)$ at least to the ϱ_rth power. TATE ([*27*], substantially repeated in [*28*]) has conjectured that this is the exact power – in other words, that every factor $(1-q^r x)$ arises from a rational r-dimensional cycle. This can be rephrased in terms of the zeta-function (6.1) as follows:

Conjecture D. *With the notation above, the order of the pole of $Z_V(s)$ at $s=r$ is equal to the rank of the group of classes of r-cycles defined over k on V, modulo algebraic equivalence.*

Tate has verified this for a number of special varieties. For the case when $d=2$, so that V is a surface, see below.

Clearly there can be no analogous statement when i is odd. It is known that $P_1(x)$ depends only on the Albanese variety of V; and assuming the existence of Weil's 'higher Jacobians', some of the factors of $P_{2r+1}(x)$ must come from the higher Jacobian of V in dimension r. (One possible definition of the higher Jacobian is as follows. Let W be a variety which parametrizes some maximal family of r-dimensional subvarieties of V; then the higher Jacobian in dimension r is that Abelian variety which is universal for maps from any such W to any Abelian variety. Perhaps for good enough V it is even the Albanese variety of each such W.) For the special case of the cubic three-fold, see [4]. Such scanty evidence as exists suggests that all those factors of $P_{2r+1}(x)$ for which q^r divides $\alpha_{2r+1,j}$ arise in this way.

More generally, for any i and r with $r \leqslant \frac{1}{2}i \leqslant d$ we can pick out those factors $(1-\alpha_{ij}x)$ of $P_i(x)$ for which q^r divides α_{ij}. Is it true that these and only these factors arise from cohomology classes which are in some sense built up from r-dimensional cycles on V – not necessarily defined over k?

In the special case where V is a surface, ARTIN and TATE [29] have gone much further. Let NS denote the Néron-Severi group of V over k – that is, the group of classes of divisors defined over k on V, modulo algebraic equivalence; let ϱ be the rank of NS and let $D_1, \ldots, D_\varrho$ be a base for NS modulo torsion. Write

$$\alpha = p_g(V) - \delta(V) \geqslant 0,$$

where p_g is the geometric genus of V and $\delta(V)$ is the "defect of smoothness" of the Picard scheme of V over k. Let $\mathrm{Br}(V)$ be the Brauer group of V over k. For details of this see GROTENDIECK [9], [10] and [11]; the Brauer group is related to the Tate-Šafarevič group, but has the advantage that for many V it can be proved to be finite and even computed. The conjecture of ARTIN and TATE is as follows:

Conjecture E. *With the notation above,*

$$P_2(q^{-s}) \sim \frac{[\mathrm{Br}(V)]\,|\det\{D_i \cdot D_j\}|}{q^\alpha [\mathrm{NS}_{\mathrm{tors}}]^2} (1 - q^{1-s})^\varrho \quad as \quad s \to 1,$$

where the curly brackets denote intersection multiplicity.

This is closely related to the variant of Conjecture C in which κ is a finitely generated field of transcendence degree 1 over a finite field. For the full details see [29]; here we only sketch the idea. Let C, defined over $\kappa = k(t)$, be a generic member of a pencil of curves on V. The zeta-function of C over κ is closely connected with that of V over k, because C is a generic fibre of V. The group of divisor classes defined over κ on C, modulo linear equivalence, is isomorphic to the Néron-Severi group of V over k; this fact lies at the heart of Néron's proof of the Néron-Severi theorem – see for example LANG [14], Chapter V. It was pointed out at the end of § 5 that Conjecture C could be expressed entirely in terms of the curve C, without overt reference to its Jacobian A. The height, on C, is a bilinear form on the group of divisor classes on C defined over κ; and by the isomorphism above it becomes a bilinear form on NS. Since there is already one such form given by intersection number, these two ought to be the same; and this is confirmed by a detailed analysis.

The status of Conjecture E is much better than that of the previous conjectures, for ARTIN and TATE have proved that at least its non-p part follows from statements about V which are apparently much weaker. The precise result they prove is as follows:

Theorem. *For given V and k suppose that either*

(i) *the l-primary part of* $\mathrm{Br}(V)$ *is finite for some prime $l \neq p$; or*

(ii) *$P_2(q^{-s})$ has a zero of order precisely ϱ at $s=1$.*

Then the non-p *part of* $\mathrm{Br}(V)$ *is finite and*

$$P_2(q^{-s}) \sim \frac{[\mathrm{Br}(V)_{\mathrm{non}-p}]\,|\det\{D_i \cdot D_j\}|}{p^{\nu}[\mathrm{NS}_{\mathrm{tors}}]^2}(1-q^{1-s})^{\varrho} \quad \text{as} \quad s \to 1$$

for some integer ν.

Tate has proved that $\mathrm{Br}(V)$ is finite in a number of cases, in particular when V is a product of two curves. This last result is closely connected with his proof that if two Abelian varieties defined over a finite field have the same zeta-function, then they are isogenous.

7. Varieties over algebraic number fields

Let V be a complete non-singular variety of dimension d defined over an algebraic number field κ. For almost all finite primes $\mathfrak{p}$ of κ, V has a non-singular reduction $V_{\mathfrak{p}}$ modulo $\mathfrak{p}$; and if q is the absolute norm of $\mathfrak{p}$ then $V_{\mathfrak{p}}$ has a local zeta-function given by (6.1) and (6.2). For each i with $0 \leqslant i \leqslant 2d$, we can associate with V an L-series

$$L_i(V,s) = \prod_{\mathfrak{p}} \{P_i(q^{-s})\}^{-1} = \prod_{\mathfrak{p}} \{\prod_j (1-\alpha_{ij} q^{-s})\}^{-1}, \tag{7.1}$$

where the outer product is taken over all $\mathfrak{p}$ for which V has a good reduction. (This is the process which we applied to elliptic curves in § 2, and implicitly to arbitrary curves in § 5. However, in contrast with these cases, we shall not attempt here to supply any factors corresponding to the bad finite primes or the infinite primes.) Even assuming (6.3), the product (7.1) only converges in $\Re s > 1 + \frac{1}{2}i$, except of course in the trivial case when $B_i = 0$. However, it is generally believed that $L_i(s)$ can be analytically continued through the entire complex plane, and that there is a functional equation connecting $L_i(s)$ with $L_i(1+i-s)$.

The cases $i=0$ and 1 (and by symmetry $i=2d-1$ and $2d$) need no further discussion. $L_0(s)$ is, except for some missing factors, the classical Riemann zeta-function of κ; and it is well known that its behaviour near $s=1$ (the real point on the boundary of the half-plane of convergence) gives valuable information about κ. $L_1(s)$ depends only on the Albanese variety A of V, and is in fact the $L^*(A, s)$ of § 5 with some factors missing; according to Conjecture C its most interesting behaviour is at $s=1$, which is the real point a distance $\frac{1}{2}$ from the half-plane of convergence, and is the centre of the critical strip. As in § 6, we must therefore expect a fundamental difference between odd and even values of i.

When $i=2r$ is even, TATE [27] has suggested that the analogue of Conjecture D still holds, in the following form:

Conjecture F. *With the notation above, the order of the pole of* $L_{2r}(V, s)$ *at* $s=r+1$ *is equal to the rank of the group of classes of r-cycles defined over* κ *on* V, *modulo algebraic equivalence.*

It should be emphasized that the heuristic deduction of this from Conjecture D along the lines "$P_i(q^{-s}) \sim C_\mathfrak{p}(1-q^{r-s})^\varrho$ for some constants $C_\mathfrak{p}$; hence $L_i(V, s) \sim C\{\zeta_\kappa(s-r)\}^\varrho \sim C'(s-r-1)^{-\varrho}$ near $s=r+1$" is totally misleading; it is not even true that V and $V_\mathfrak{p}$ have the same ϱ_r for almost all $\mathfrak{p}$. The simplest counter-example is $V=\Gamma\times\Gamma$, where Γ is an elliptic curve which does not admit complex multiplication; here $\varrho_1=3$ for V, but $\varrho_1=4$ for almost all $V_\mathfrak{p}$. There is strong reason to suppose that any sufficiently general quartic surface is also a counter-example.

Tate has verified this conjecture in a few special cases. Moreover, by taking V to be the d-fold product of an elliptic curve Γ not admitting complex multiplication with itself, he has deduced from it some very interesting results on the distribution of the arg α_p as p varies. (Here α_p is defined by (2.1).) These results agree with the numerical evidence. The constant

$$\lim (s-r-1)^\varrho L_{2r}(V, s)$$

has been evaluated for a number of varieties V. The results suggest that the constant is significant, but as yet there is nothing with which it can be even conjecturally identified.

When $i=2r+1$ is odd almost nothing is known, except of course for the case $i=1$ treated in § 5 and the symmetric case $i=2d-1$. There is however one striking and wholly unexplained phenomenon. Suppose that $V=\Gamma\times\Gamma\times\Gamma$, where Γ is an elliptic curve; for convenience of notation we assume that Γ is defined over **Q** and its local zeta-function is given by (2.1). Now

$$P_3(p^{-s})=(1-\alpha_p^3 p^{-s})(1-\bar{\alpha}_p^3 p^{-s})(1-\alpha_p p^{1-s})^9(1-\bar{\alpha}_p p^{1-s})^9$$

and so $L_3(V,s)=L_3^0(\Gamma,s)\,\{L_1(\Gamma,s-1)\}^9$ say, where

$$L_3^0(\Gamma,s)=\prod\{(1-\alpha_p^3 p^{-s})(1-\bar{\alpha}_p^3 p^{-s})\}^{-1}.$$

It is reasonable to ascribe the term $\{L_1(\Gamma,s-1)\}^9$ to the higher Jacobian in dimension 1; and in any case its behaviour near $s=2$ is completely described by Conjecture B. We are therefore led to examine $L_3^0(\Gamma,s)$ near $s=2$.

Now suppose that Γ has the form

$$y^2=x^3-Dx. \tag{7.2}$$

Using the ideas of § 3, though in a more complicated form, it is possible to give a closed expression for $L_3^0(\Gamma,2)$ in terms of division values of elliptic functions; and this expression has been evaluated for about 100 values of D. The structure of the results closely resembles what we would be able to say about $L_1(\Gamma,1)$ if we had only the results of the calculation, and lacked any number-theoretic theory of elliptic curves to attach them to. To clarify this, I formulate a weak version of Conjecture A for the curve (7.2). For this purpose we employ "fudge factors" λ_∞ and λ_p for each prime p dividing $2D$. These have explicit definitions in terms of the local properties of D, which are too long to give here; but it should be noted that $\lambda_p=1$ or 2 for each finite p. They are in fact just the $\{M_p(\Gamma)\}^{-1}$, possibly multiplied by rational squares; but we are forbidden to use that interpretation in the present context. Write

$$\mu=L_1(\Gamma,1)\lambda_\infty\prod\lambda_p$$

the "fudged" value of $L_1(\Gamma,1)$.

Weak Conjecture A. *The rational integer μ is a perfect square. By local considerations involving only the sign of D and those primes which divide $2D$, we can state a sufficient condition for $\mu=0$; but this condition is not a necessary one.*

That μ is a rational integer can be proved from the explicit formula.

That μ is a perfect square is equivalent to saying that [Ш] is a perfect square. The sufficient condition for $\mu=0$ is just the necessary and sufficient condition for (7.2) to have an odd number of first descents, and so for g to be odd; this is sufficient, but not necessary, for $g>0$.

A precisely analogous conjecture fits the numerical evidence for $L_3^0(\Gamma, 2)$. We can define local "fudge factors" λ'_∞ and λ'_p; these are not the same as λ_∞ and λ_p above, but we do have $\lambda'_p=1$ or 2. Write

$$\mu' = L_3^0(\Gamma, 2)\,\lambda'_\infty \prod \lambda'_p.$$

It can be proved that μ' is a rational integer; and the numerical evidence shows that μ' must always be a square. (There is no possibility of coincidence, for the numbers involved are much larger than with μ.) Moreover one can state local conditions (not the same as for μ) which appear to be sufficient but not necessary for $\mu'=0$. Presumably there is a duality theory and a descent argument of some sort which underlies all this; but what it is I have no idea.

References

[1] BIRCH, B. J., and N. M. STEPHENS: The parity of the rank of the Mordell-Weil group. Topology 5, 295–299 (1966).

[2] BIRCH, B. J., and H. P. F. SWINNERTON-DYER: Notes on elliptic curves I, J. reine angew. Math. **212**, 7–25 (1963).

[3] BIRCH, B. J., and H. P. F. SWINNERTON-DYER: Notes on elliptic curves II, J. reine angew. Math. **218**, 79–108 (1965).

[4] BOMBIERI, E. and H. P. F. SWINNERTON-DYER: The zeta function of a cubic threefold. Annali di Pisa **21**, 1–29 (1967).

[5] CASSELS, J. W. S.: Arithmetic on curves of genus 1 (IV) Proof of the *Hauptvermutung*. J. reine angew. Math. **211**, 95–112 (1962).

[6] CASSELS, J. W. S.: Diophantine equations with special reference to elliptic curves, J. Lond. Math. Soc. **41** ,193–219 (1966).

[7] EICHLER, M.: Quaternäre quadratische Formen und die Riemann Vermutung für die Kongruenzzetafunktion. Arch. Math. **5**, 355–366 (1954).

[8] FRICKE, R.: Die elliptischen Funktionen und ihre Anwendungen. Vol. **2**, Leipzig 1922.

[9] GROTHENDIECK, A.: Le groupe de Brauer I. Sém. Bourbaki **290** (1965).

[10] GROTHENDIECK, A.: Le groupe de Brauer II. Sém. Bourbaki **297** (1965).

[11] GROTHENDIECK, A.: Le groupe de Brauer III. Mimeo notes I.H.E.S. (1966).

[12] HECKE, E.: Eine neue Art von Zetafunktionen und ihre Beziehungen zur Verteilung der Primzahlen. Zweite Mitteilung. Math. Z. **6**, 11–51 (1920).

[13] HECKE, E.: Über Modulfunktionen und die Dirichletschen Reihen mit Eulerschen Produktentwicklung. Math. Ann. **114**, 1–28 and 316–351, (1937).

[14] LANG, S.: Diophantine Geometry. New York 1962.

[15] LANG, S.: Les formes bilinéaires de Néron et Tate. Sém. Bourbaki **274** (1964).

[16] MORDELL, L. J.: On the rational solution of the indeterminate equations of the third and fourth degrees. Proc. Camb. Phil. Soc. **21**, 179–192, (1922).

[17] NÉRON, A.: Modèles minimaux des variétés abéliennes sur les corps locaux et globaux. Publ. IHES **21** (1964).

[18] NÉRON, A.: Quasi-fonctions et hauteurs sur les variétés abéliennes. Ann. Math. **82**, 249–331, (1965).

[19] OGG, A. P.: Elliptic curves and wild ramification, Amer. J. Math. **89**, 1–21 (1967).

[20] OGG, A. P.: Abelian curves of 2-power conductor. Proc. Camb. Phil. Soc. **62**, 143–148, (1966).

[21] PETERSSON, H.: Konstruktion der sämtlichen Lösungen einer Riemannschen Funktionalgleichung durch Dirichlet-Reihen mit Eulerscher Produktentwicklung. Math. Ann. **116**, 401–412 (1939) and **117**, 39–64 and 277–300, (1940).

[22] SHIMURA, G. and TANIYAMA, Y.: Complex multiplication of Abelian varieties. Publ. Math. Soc. Japan **6** (1961).

[23] SHIMURA, G.: Correspondances modulaires et les fonctions zeta de courbes algébriques, J. Math. Soc. Japan **10**, 3–28, (1958).

[24] STEPHENS, N. M.: Thesis Manchester, (1965).

[25] STEPHENS, N. M.: Conjectures concerning elliptic curves. (In press).

[26] TATE, J.: Duality theorems in galois cohomology over number fields. Proc. Intern. Congress Math. Stockholm, 288–295, (1962).

[27] TATE, J.: Algebraic cohomology classes, Mimeo notes Woods Hole (1964).

[28] TATE, J.: Algebraic cycles and poles of zeta functions. Proc. Purdue Univ. Conf. 1963. New York 93–110, (1965).

[29] TATE, J.: On the conjectures of Birch and Swinnerton-Dyer and a geometric analog. Sém. Bourbaki **306** (1966).

[30] WEIL, A.: L'arithmétique sur les courbes algébriques, Acta Math. **52**, 281–315, (1928).

[31] WEIL, A.: On algebraic groups and homogenous spaces, Amer. J. Math. **77**, 493–512, (1955).

[32] WEIL, A.: Über die Bestimmung Dirichletscher Reihen durch Funktionalgleichungen, Math. Ann. **168**, 149–156 (1967).

p-Divisible Groups [1]

J. T. Tate

Introduction

After a brief review of facts about finite locally free commutative group schemes in § 1, we define p-divisible groups in § 2, and discuss their relation to formal Lie groups. The § 3 contains some theorems about the action of $\mathrm{Gal}(\bar{K}/K)$ on the completion C of the algebraic closure $\bar{K}$ of a local field K of characteristic 0. In § 4 these theorems are applied to obtain information about the Galois module of points of finite order on a p-divisible group G defined over the ring of integers R in such a field K, and to prove that G is determined by that Galois module, or, what is the same, by its generic fiber $G \times_R K$.

The notion of p-divisible group and the basic theorems of § 2 are the result of joint work with Serre, and § 3 and § 4 owe much to discussions with him. Although p-divisible groups are interesting enough in their own right, our main motivation for studying them has been their applications to abelian varieties. For some of these, and for further results on p-divisible groups, as well as additional bibliography, see Serre, [*10*] and [*11*].

This text owes much – probably its very existence – to the efforts of T. Springer, who wrote a first draft shortly after the conference. In several places, and particularly in § 3, he has made improvements on the original oral exposition. I thank him heartily for his help.

§ 1. Finite group schemes

(1.1). Let R be a commutative ring (or a prescheme). By a *finite group of order m over R* we shall mean a group scheme G which is locally free of rank m over R. Such a G is defined by a locally free (sheaf of) algebra(s) A of rank m over R. The group structure is described by homomorphisms

[1] The research described here has been partially supported by NSF.

$\mu: A \to A \otimes A$, $\varepsilon: A \to R$, and an automorphism $i: A \to A$, describing multiplication, neutral element, and inverse, respectively. These are subject to a number of rather obvious conditions (see for example [4] or [6]).

Examples. (a) Let Γ be a finite abstract group of order m. Let A be the ring of R-valued functions on Γ, let $(\mu f)(s, t) = f(st)$, let $(if)(s) = f(s^{-1})$, and let $\varepsilon f = f(1)$. Then $\Gamma = \mathrm{Spec}(A)$ is a finite group of order m over R.

(b) Let $A = R[X]/(X^m - 1)$, with $\mu(x) = x \otimes x$, where x is the image of X in A. Then $\mathrm{Spec}(A)$ is a finite group of order m over R which is denoted by $\boldsymbol{\mu}_m$; it is the kernel of the homomorphism m: $\mathbf{G}_m \to \mathbf{G}_m$, where $\mathbf{G}_m$ denotes the "multiplicative group", viewed over R.

(c) Let $a, b \in R$ with $ab = 2$. Put $A = R + Rx$, with $x^2 + ax = 0$, and put $\mu x = x \otimes 1 + 1 \otimes x + bx \otimes x$. Then $G_{a,b} = \mathrm{Spec}(A)$ is the most general group of order 2 over R whose affine algebra A is free over R. Moreover, $G_{a,b}$ and $G_{a',b'}$ are isomorphic if and only if there is a unit u in R such that $a' = ua$, $b' = u^{-1}b$.

(1.2). Duality

From now on we assume all groups to be commutative. Then there is a duality theory, due to Cartier (see [3], p. 106). Let $G = \mathrm{Spec}\, A$ be a finite group over R and $m: A \otimes A \to A$ define the multiplication in A and $\mu: A \to A \otimes A$ the group law. Put $A' = \mathrm{Hom}_{R\text{-modules}}(A, R)$. We then get dual homomorphisms

$$\mu': A' \otimes A' \to A', \quad m': A' \to A' \otimes A'.$$

μ' defines an algebra structure on A' and it is easy to check that m' defines a product on $G' = \mathrm{Spec}(A')$, which makes it into a group scheme. The order of G' is equal to that of G. G' is called the *Cartier dual* of G. There is a canonical isomorphism $G \tilde{\to} (G')'$.

If T is a prescheme over R, then the group $G'(T)$ is canonically isomorphic to $\mathrm{Hom}_T(G, \mathbf{G}_m)$ (homomorphisms of group schemes over T of $G \times_R T$ into $\mathbf{G}_m \times T$, $\mathbf{G}_m$ denoting the multiplicative group).

Examples. (a) Let $G = \mathbf{Z}/m\mathbf{Z}$ be the cyclic group of order n (as in Example (a) of (1.1)). Then $G' = \boldsymbol{\mu}_m$.
(b) If $G = G_{a,b}$ (Example (c) of (1.1)), then $G' = G_{b,a}$.

(1.3). Short exact sequences

A sequence

$$0 \to G' \xrightarrow{i} G \xrightarrow{j} G'' \to 0 \tag{1}$$

of finite R-groups is called *exact* if i is a closed immersion which identifies G' with the kernel of j (in the sense of categories), whereas j is faithfully flat. If j is given, it is easy to get G' (it is the inverse image of the unit section of G''). Given i, one can construct G''. This is more delicate (see [*4*], Exposé V and VI_B or [*7*]).

If (1) is exact, then we have for the orders m, m', m'' of G, G', G'' the relation $m = m'm''$. This follows from the fact (proved loc. cit.) that the graph of the equivalence relation in $G \times_R G$ defined by G' is isomorphic to $G \times_{G''} G$.

Finally, the dual of an exact sequence (1) is also exact.

(1.4). Connected and étale groups

In this section we suppose R is a *complete noetherian local* ring. If G is a group of finite order over R, there is a canonical exact sequence

$$0 \to G^0 \xrightarrow{i} G \xrightarrow{j} G^{et} \to 0,$$

where G^0 is *connected* and G^{et} is *étale* over R. If the corresponding affine rings are A^0, A, and A^{et}, then A is a product of local R-algebras, and A^0 is the local quotient of A through which the map $\varepsilon: A \to R$ factors, while A^{et} is the maximal étale subalgebra of A. The map i is an open and closed immersion, and G^0 is the maximal connected subgroup of G. For varying G, the functors $G \mapsto G^0$, and $G \mapsto G^{et}$ are *exact*.

G is *connected* if $G^0 = G$. In that case, the order of G is a power of the characteristic exponent of the residue field k of R, i.e., is 1 if $\mathrm{char}(k) = 0$, and is a power of p if $\mathrm{char}\, k = p > 0$. This follows from the theory of finite group schemes over k (or even over $\bar{k}$).

G is *étale* if $G = G^{et}$. The functor $G \mapsto G(\bar{k})$ is an equivalence of the category of étale R-groups of finite order and finite π-modules on which π operates continuously, where $\pi = \mathrm{Gal}(\bar{k}/k)$ is the *fundamental group* of R. Given such a π-module Γ, the étale finite R-group Γ corresponding to it is given by $\Gamma = \mathrm{Spec}\, A$, where A is the ring of all functions $\Gamma \to R_{et}$ which commute with π, and where R_{et} is a "maximal local étale integral extension" of R (i.e., a maximal unramified extension of R in the old terminology, if R is a complete discrete valuation ring), and where π operates on R_{et} in the unique way compatible with its operation on the residue field extension k_{et}/k (the residue field k_{et} of R_{et} being a separable algebraic closure of k). For general (not necessarily étale) G, we have $G^{et} = G(\bar{k})$.

§ 2. p-divisible groups

(2.1). Definition

Let p be a prime number, and h an integer $\geqslant 0$. A *p-divisible group G over R of height h* is an inductive system

$$G = (G_\nu, i_\nu), \qquad \nu \geqslant 0,$$

where

(i) G_ν is a finite group scheme over R of order $p^{\nu h}$,

(ii) for each $\nu \geqslant 0$,

$$0 \to G_\nu \xrightarrow{i_\nu} G_{\nu+1} \xrightarrow{p^\nu} G_{\nu+1}$$

is exact (i.e., G_ν can be identified via i_ν with the kernel of multiplication by p^ν in $G_{\nu+1}$).

These axioms for ordinary abelian groups would imply

$$G_\nu \cong (\mathbf{Z}/p^\nu\mathbf{Z})^h \quad \text{and} \quad G = \varinjlim G_\nu = (\mathbf{Q}_p/\mathbf{Z}_p)^h .$$

A *homomorphism* $f: G \to H$ of p-divisible groups is defined in the obvious way: if $G = (G_\nu, i_\nu)$, $H = (H_\nu, i_\nu)$ then f is a system of homomorphisms $f_\nu: G_\nu \to H_\nu$ of R-groups, which is such that $i_\nu \cdot f_\nu = f_{\nu+1} \cdot i_\nu$ for all $\nu \geqslant 1$.

By iteration, one gets from the i_ν closed immersions

$$i_{\nu,\mu}: G_\nu \to G_{\mu+\nu}$$

for all $\mu, \nu \geqslant 0$, which identify G_ν with the kernel of multiplication by p^ν in all $G_{\mu+\nu}$. It follows that the homomorphism

$$p^\mu: G_{\mu+\nu} \to G_{\mu+\nu}$$

can be factored through G_ν and then a consideration of orders shows that we have an exact sequence

$$0 \to G_\mu \xrightarrow{i_{\mu,\nu}} G_{\mu+\nu} \xrightarrow{j_{\mu,\nu}} G_\nu \to 0 , \tag{2}$$

where $j_{\mu,\nu}$ is the unique homomorphism such that $i_{\nu,\mu} \cdot j_{\mu,\nu} = p^\mu$.

Examples. (a) Let X be an abelian scheme over R of dimension d. Let X_n be the kernel of multiplication by n. Then (X_{p^ν}, i_ν) (i_ν denoting the obvious inclusion) is a p-divisible group $X(p)$, with height $h = 2d$.

(b) The same construction can be performed for other groups over R. If one takes $X = \mathbf{G}_m$, the resulting p-divisible group is $\mathbf{G}_m(p) = (\boldsymbol{\mu}_{p^\nu}, i_\nu)$. Its height is 1.

Question: Are there any p-divisible groups over $\mathbf{Z}$ other than products of powers of $\mathbf{G}_m(p)$ and of $\mathbf{Q}_p/\mathbf{Z}_p$?

(2.2). Relations with formal Lie groups

In this section we assume R complete, noetherian, local, with residue field k of characteristic $p>0$. For our present purposes, an *n-dimensional formal Lie group Γ over R* can be defined as a suitable homomorphism of the ring $\mathscr{A}=R[[X_1,\cdots,X_n]]$ of formal power series over R in n variables X_i into $\mathscr{A}\hat{\otimes}_R\mathscr{A}$, the ring of formal power series in $2n$ variables Y_i, Z_j. Such a homomorphism can be described by a family $f(Y,Z)=(f_i(Y,Z))$, of n power series in $2n$ variables, f_i being the image of X_i.

The following axioms are imposed

(i) $X=f(X,o)=f(o,X)$,

(ii) $f(X,f(Y,Z))=f(f(X,Y),Z)$,

(iii) $f(X,Y)=f(Y,X)$, (since we consider only commutative groups).

We write $X*Y=f(X,Y)$. It follows from the axioms that $X*Y=X+Y+$ terms in higher powers of the variables.

Put $\psi(X)=X*\cdots*X$ (p times). This determines a homomorphism $\psi:\mathscr{A}\to\mathscr{A}$, which corresponds to multiplication by p in Γ. Γ is said to be *divisible* if $p:\Gamma\to\Gamma$ is an isogeny. This means that ψ makes $\mathscr{A}$ into a *free* module of finite rank over itself.

We can then repeat the construction of Example (a) in (2.1), obtaining a p-divisible group $\Gamma(p)=(\Gamma_{p^\nu},i_\nu)$ of height h over R, where p^h is the degree of the isogeny $p:\Gamma\to\Gamma$. (This degree is a power of p because it is equal to the order of the finite R-group $\Gamma_p=\operatorname{Ker}p$, which is connected (see 1.4).) More generally, for arbitrary ν, we have $(\Gamma(p))_\nu=\Gamma_{p^\nu}=\operatorname{Spec}A_\nu$, where $A_\nu=\mathscr{A}/J_\nu$, and where $J_\nu=\psi^\nu(I)\mathscr{A}$ is the ideal in $\mathscr{A}$ generated by the elements $\psi^\nu(X_i)$, $1\leqslant i\leqslant n$. (Here $I=J_0$ denotes the ideal generated by the variables X_i). Clearly, each A_ν is a local ring; hence $\Gamma(p)$ is a *connected* p-divisible group (i.e., each $\Gamma(p)_\nu$ is connected).

Proposition 1. *Let R be a complete noetherian local ring whose residue field k is of characteristic $p>0$. Then $\Gamma\mapsto\Gamma(p)$ is an equivalence between the category of divisible commutative formal Lie groups over R and the category of connected p-divisible groups over R.*

Remark. We can only sketch the proof here, omitting many technical details. The techniques involved are amply covered in Gabriel's exposés VII_A and VII_B of [*4*].

Let Γ be a divisible formal group over R. Let $\mathfrak{m}$ be the maximal ideal of R so that, with the previous notations, $\mathfrak{m}\mathscr{A}+I=M$, say, is the maximal ideal of $\mathscr{A}$.

Lemma 0. *The ideals* $\mathfrak{m}^\nu\mathscr{A}+J_\nu$ *constitute a fundamental system of neighborhoods of* 0 *in the* M*-adic topology of* $\mathscr{A}$.

Indeed, $\mathscr{A}/(\mathfrak{m}^\nu A+J_\nu)=A_\nu/\mathfrak{m}^\nu A_\nu$ is an Artin ring, so the ideals in question are M-adically open. On the other hand, they are arbitrarily small, because we have

$$\psi(X_i) = pX_i + (\text{terms of degree} \geqslant 2),$$

hence $\psi(I)\subset pI+I^2\subset(\mathfrak{m}\mathscr{A}+I)\,I=MI$, and consequently, by induction on ν, we have $J_\nu=\psi^\nu(I)\,\mathscr{A}\subset M^\nu I$.

From Lemma 0 it follows that the map $\mathscr{A}\to\varprojlim(\mathscr{A}/J_\nu)=\varprojlim(A_\nu)$ is bijective, because $\mathscr{A}$ is M-adically complete. From this bijectivity it is easy to see that the functor $\Gamma\to\Gamma(p)$ is fully faithful.

To complete the proof of Prop. 1 we must show that any given connected p-divisible group G over R is isomorphic to some $\Gamma(p)$. Let $G=(G_\nu, i_\nu)$ and $G_\nu=\operatorname{Spec}(A_\nu)$. The inclusions $i_\nu\colon G_\nu\to G_{\nu+1}$ make (A_ν) into a projective system. Put $A=\varprojlim A_\nu$. The group law on G defines a homomorphism $A\to A\widehat{\otimes}_R A$, which will have the required properties for a formal Lie group, once we know that A is isomorphic to $R[[X_1,\cdots,X_n]]$ for some n. Then it also follows easily that the formal Lie group Γ is divisible and that G is isomorphic to $\Gamma(p)$.

To prove that A is an algebra of formal power series over R, one first observes that A is flat over R. (Indeed the R-module A is isomorphic to a countable direct product of copies of R, because the A_ν are free of finite type over R, and the maps $A_{\nu+1}\to A_\nu$ surjective.) This being so, one is reduced by standard procedures to the case in which $R=k$ is a field of characteristic $p>0$. In that case, the injective limits of finite commutative group schemes of p-power order over k form an abelian category, and a p-divisible group over k is just an object in that category on which p is surjective with finite kernel. There is an exact functor $G\mapsto G^{(p)}$ from that category to itself which preserves the order of the finite objects, and there are homomorphisms of functors F (Frobenius) and V (Verschiebung) such that the following diagram is commutative for any G in the category:

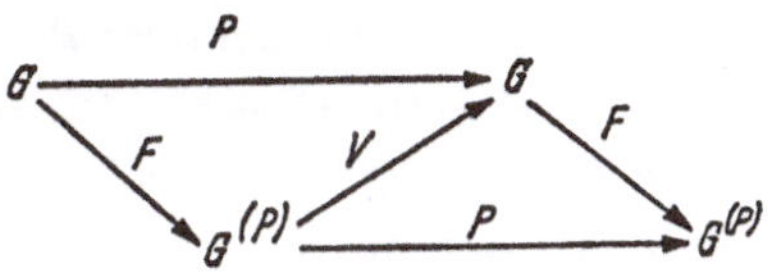

(See [3], p. 98, or [4], exposé VII, or [6], p. 18, 19.) If G is p-divisible of height h, then so is $G^{(p)}$, and the diagram shows then that F and V are surjective, with finite kernels of order $\leqslant p^h$.

Suppose now that G is the connected p-divisible group over k we were considering before this general discussion. For each v, let H_v be the kernel of $F^v: G \to G^{(p^v)}$. Then we have $H_v \subset G_v$, and also $G_v \subset H_N$ for sufficiently large N (depending on v), because G_v is finite and connected. Thus, we have $A = \varprojlim A_v = \varprojlim B_v$, where $H_v = \operatorname{Spec} B_v$. Let I_v be the augmentation (i.e., the maximal) ideal of B_v, let $I = \varprojlim I_v$ be the augmentation ideal of A, and let $x_1, \cdots, x_n$ be elements of I whose images form a k-base for I_1/I_1^2. Then the images of the x_i in B_v generate I_v for each v, because $I_v/I_v^2 \to I_1/I_1^2$ is bijective. (Since H_1 is the kernel of F in H_v, the kernel of $I_v \to I_1$ is generated by the p-th powers of the elements of I_v, so that kernel is in I_v^2.) Now consider the homomorphisms $u_v: k[X_1, \cdots, X_n] \to B_v$ which send X_i to the image of x_i in B_v. These are surjective, by the above; on the other hand, $\operatorname{Ker} u_v$ contains the elements $X_i^{p^v}$ because F^v kills H_v. But $\operatorname{rank}(B_v) = (\operatorname{rank}(B_1))^v$, since F is *surjective*, and $\operatorname{rank}(B_1) = p^n$, by the structure theory of finite groups killed by F. Since the ideal $(X_1^{p^v}, \cdots, X_n^{p^v})$ is of codimension p^{nv} in $k[X]$, it follows that that ideal is the kernel of u_v, and hence that the n_v induce an isomorphism $u: k[[X_1, \cdots, X_n]] \tilde{\to} A$. This completes our sketch of the proof of Proposition 1.

If $G = (G_v, i_v)$ is now any p-divisible group over our complete noetherian local R, the connected components G_v^0 determine a *connected* p-divisible group G^0. From the exact sequences

$$0 \to G_v^0 \to G_v \to G_v^{et} \to 0$$

one gets an exact sequence

$$0 \to G^0 \to G \to G^{et} \to 0, \tag{4}$$

where G^{et} is an *étale* p-divisible group. The dimension, n, of the formal Lie group corresponding to G^0 is, by definition, the *dimension* of G.

Proposition 2. *The discriminant ideal of A_v over R is generated by $p^{nvp^{hv}}$, where $h = ht(G)$ and $n = \dim(G)$.*

In general, for a finite group $H = \operatorname{Spec} B$ over R, let $\operatorname{disc}(H)$ denote the discriminant ideal of B over R. If $0 \to H' \to H \to H'' \to 0$ is an exact sequence of finite groups over R (cf. 1.3) of orders m', m, and m'', then by the transitivity of discriminants one proves easily that $\operatorname{disc}(H) = (\operatorname{disc} H')^{m''}$

$(\operatorname{disc} H'')^{m'}$. This fact, together with the fact that $\operatorname{disc}(H)=1$ if H is étale, allows us, via (4), to reduce the case of an arbitrary G to that of a connected $G \simeq \Gamma(p)$. As in the proof of Prop. 1 above we have then $A_\nu = \mathscr{A}/J_\nu$. Consider $\mathscr{A}$ as a free module of rank $p^{\nu h}$ over itself by means of $\varphi=\psi^\nu$. We change the notation, and consider $\mathscr{A}$ as an algebra (via φ) over another copy $\mathscr{A}'$ of $\mathscr{A}$. Denoting by I' the augmentation ideal in $\mathscr{A}'$ (generated by the X_i'), we have $A_\nu \simeq \mathscr{A}/I'\mathscr{A}$, so it suffices to prove that the discriminant ideal of $\mathscr{A}$ over $\mathscr{A}'$ is generated by the desired power of p.

To do this, consider the modules of formal differentials Ω and Ω' of $\mathscr{A}$, resp. $\mathscr{A}'$. They are free modules over $\mathscr{A}$ (resp. $\mathscr{A}'$) generated by the differentials of the variables dX_i, resp. dX_i', $1 \leqslant i \leqslant n$. The homomorphism $\varphi: \mathscr{A}' \to \mathscr{A}$ induces an $\mathscr{A}'$-linear map $d\varphi: \Omega' \to \Omega$. Choosing bases in Ω, resp. Ω', we get a basis element θ, resp. θ', of $\Lambda^n \Omega$, resp. $\Lambda^n \Omega'$. Let $d\varphi(\theta') = a\theta$, with $a \in \mathscr{A}$. Then one has

Lemma 1. *The discriminant ideal of $\mathscr{A}$ over $\mathscr{A}'$ is generated by $N_{\mathscr{A}/\mathscr{A}'}(a)$.*

Granting this lemma, the proof of Prop. 2 is finished as follows. Choose a basis (ω_i) of Ω consisting of translation-invariant differentials, i.e., differentials such that if $\mu: \mathscr{A} \to \mathscr{A} \hat{\otimes}_R \mathscr{A}$ defines the formal group structure $d\mu: \Omega \to \Omega \oplus \Omega$ satisfies $d\mu(\omega_i) = \omega_i \oplus \omega_i$. Using the corresponding basis (ω_i') in the copy Ω' of Ω we have, by the definition of φ as arising from the homomorphism $p^\nu: \Gamma \to \Gamma$, that $d\varphi(\omega_i') = p^\nu \omega_i$, whence $a = p^{\nu n}$. Proposition 2 then follows by Lemma 1.

As to a proof of that lemma, it can be based on the existence of a trace map $\operatorname{Tr}: \Lambda^n \Omega \to \Lambda^n \Omega'$ with the following properties:

(i) Tr is $\mathscr{A}'$-linear.

(ii) The map $a \mapsto (\theta \mapsto \operatorname{Tr}(a\theta))$ establishes an isomorphism of $\mathscr{A}'$-modules

$$\mathscr{A} \xrightarrow{\sim} \operatorname{Hom}_{\mathscr{A}'}(\Lambda^n \Omega, \Lambda^n \Omega').$$

(iii) If $\theta \in \Omega'$ and $a \in \mathscr{A}$, then

$$\operatorname{Tr}(a \cdot d\varphi(\theta)) = (\operatorname{Tr}_{\mathscr{A}/\mathscr{A}'}(a))\, \theta.$$

Such a trace map exists whenever $\mathscr{A} \simeq \mathscr{A}' \simeq R[[X_1, \ldots, X_n]]$ and $\varphi: \mathscr{A}' \to \mathscr{A}$ is an R-algebra homomorphism making $\mathscr{A}$ a free $\mathscr{A}'$-module of finite rank; see for example Hartshorne's notes on *Residues and Duality*, Springer lecture notes 20, 1966, at least for the corresponding "nonformal" situation.

(2.3). Duality for p-divisible groups

Let $G=(G_\nu, i_\nu)$ be a p-divisible group over R. For each ν, let G'_ν be the Cartier dual of G (cf. 1.2). The exact sequences (2) in (2.1) with $\mu=1$ shows that we have injective homomorphisms

$$i'_\nu: G'_\nu \to G'_{\nu+1},$$

where i_ν is the dual of the map $G_{\nu+1} \to G_\nu$ induced by multiplication by p. It is easy to check that the system $G'=(G'_\nu, i'_\nu)$ satisfies the axioms of (2.1), and is therefore a p-divisible group, called the *dual of* G. Clearly, G' and G have the same height. Of course, this notion of dual makes sense over any base ring (or prescheme) R. In case R is complete local noetherian with residue characteristic p, as in 2.2, so that the dimension of G is defined (cf. the lines before Prop. 2), we have the following all-important

Proposition 3. *Let n and n' be the dimension of G and its dual G'. Then $n+n'=h$, the height of G and G'.*

The dimension and height of G do not change if we reduce G mod the maximal ideal of R. Hence we are reduced at once to the case in which $R=k$, a field of characteristic p. From diagram (3) we get an exact sequence

$$0 \to \operatorname{Ker} F \to \operatorname{Ker} p \to \operatorname{Ker} V \to 0.$$

Now $\operatorname{Ker} p = G_1$ has order p^h, and $\operatorname{Ker} F$ has order p^n. (F is injective on G^{et}, so the kernel of F in G is the same as that of F in the connected component G^0. Viewing G^0 as a formal Lie group on n parameters, we see that the order of $\operatorname{Ker} F$ is p^n, as remarked in the proof of Prop. 1.) Since F and V are dual with respect to Cartier duality one checks that $\operatorname{Ker} V$ is the Cartier dual of the Cokernel of the map $F: G'_\nu \to G_1'^{(p)} = (G_1^{(p)})'$, and consequently $\operatorname{Ker} V$ has order $p^{n'}$. Now the assertion follows from the multiplicativity of orders in an exact sequence.

Examples. a) The p-divisible group $\mathbf{G}_m(p)$ has $h=n=1$, and is dual to the étale p-divisible group $\mathbf{Q}_p/\mathbf{Z}_p$ which has $h=1$, $n=0$.

b) Let X be an abelian scheme of dimension n over R. If the dual abelian scheme X' exists, then we have $(X(p))' \simeq X'(p)$. Both $X(p)$ and $X'(p)$ have height $2n$ and dimension n. The connected component $X(p)^0$ of $X(p)$ is the formal completion of X along the zero section, and can have any height between n and $2n$. For example, if X is an elliptic curve ($n=1$), then the height of $X(p)^0$ is 1 or 2 according as the "Hasse invariant" of $X(p)$ is non-zero or zero.

(2.4). Points; the Galois modules $\Phi(G)$ and $T(G)$

Let R be a complete discrete valuation ring, with residue field $k = R/\mathfrak{m}$ of characteristic $p > 0$, and let K be the field of fractions of R (very soon we shall assume k perfect and K of characteristic 0). Let L be the completion of a (possibly infinite) algebraic extension of K and let S be the ring of integers in L. Thus S is a complete rank 1 valuation ring, but the valuation on S may not be discrete.

Let G be a p-divisible group over R. We define the *group* $G(S)$ *of points of* G *with values in* S by

$$G(S) = \varprojlim_i G(S/\mathfrak{m}^i S),$$

where $\mathfrak{m}$ is the maximal ideal of R, and where

$$G(S/\mathfrak{m}^i S) = \varinjlim_\nu G_\nu(S/\mathfrak{m}^i S).$$

Clearly, $G(S)$ is a $\mathbf{Z}_p$-module. From the definition of p-divisible groups, $G_\nu(S/\mathfrak{m}^i S)$ is the kernel of multiplication by p^ν in $G(S/\mathfrak{m}^i S)$. Thus the kernel of multiplication by p^ν in $G(S)$ is $\varprojlim_i G_\nu(S/\mathfrak{m}^i S) \simeq G_\nu(S)$, and the torsion subgroup of $G(S)$ is given by

$$G(S)_{\text{tors}} \simeq \varinjlim_\nu G_\nu(S).$$

If G is étale over R, then the maps $G_\nu(S/\mathfrak{m}^{i+1}S) \to G_\nu(S/\mathfrak{m}^i S)$ are bijective for all i, and consequently $G(S)$ *is a torsion group if* G *is étale.*

In general, if $G_\nu = \operatorname{Spec} A_\nu$ and $A = \varprojlim_\nu A_\nu$, and $A_\nu \simeq A/J_\nu$ as in § 2, then a point $x \in G(S)$ can be identified with a homomorphism $A \to S$ which is continuous with respect to the valuation topology in S and the topology defined by the ideals $\mathfrak{m}^i A + J_\nu$ in A. In particular, if G is connected, corresponding to a formal Lie group Γ, so that $A \simeq R[[X_1, \ldots, X_n]]$, then it follows from the above remark and Lemma 0 that $G(S)$ is the group of points $x = (x_1, \ldots, x_n)$ of Γ with coordinates x_i lying in the maximal ideal of S. Thus, *if* G *is connected, then* $G(S)$ *is an analytic group over* L.

Now consider the exact sequence (4), and let A^{et} and A^0 denote the algebras of G^{et} and G^0.

Proposition 4. *If the residue field k of R is perfect, then the map $G \to G^{et}$ has a formal section, and consequently the sequence*

$$0 \to G^0(S) \to G(S) \to G^{et}(S) \to 0$$

is exact.

Proof. If $R=k$, then the exact sequence (4) splits canonically, and we have $A \simeq A^0 \hat{\otimes}_R A^{et} \simeq A^{et}[[X_1, \ldots, X_n]]$ in that case. By flatness we conclude $A \simeq A^{et}[[X]]$ in the general case.

Corollary 1. *If $x \in G(S)$, then there exists a finite extension field L' of L and an element $Y \in G(S')$, where S' is the ring of integers in L', such that $py = x$.*

Indeed, by Prop. 4, it suffices to prove this for G^{et} and G^0 separately. For G^0 it follows from the fact that the map $p: G^0 \to G^0$ makes A^0 a free A^0-module of finite rank (cf. § 2). For G^{et} we are reduced to a statement over the residue field, and the result follows because the maps $G_{\nu+1} \to G_\nu$ induced by multiplication by p are surjective.

Corollary 2. *If L is algebraically closed, then $G(S)$ is divisible.*

From now on we suppose that k is perfect and the *characteristic of K is* 0. This characteristic 0 assumption will be absolutely crucial in all that follows, because (1) there is then a logarithm map which will show that $G(S)$ is locally isomorphic to L^n, where $n = \dim G$, and (2) the $G_\nu \times_R K$ are then automatically étale, so we will know that $G_\nu(S)$ (which is isomorphic to $G_\nu(I)$ since G_ν is finite and flat over R) is isomorphic to $(\mathbf{Z}/p^\nu\mathbf{Z})^h$ for sufficiently large L (depending on ν).

The logarithm. The tangent space t_G of G at the origin is, by definition, the tangent space of the formal Lie group Γ corresponding to G^0 at the origin. We write $t_G(L)$ to denote its points with coordinates in L. Such a point is an R-linear map $\tau: A^0 \to L$ such that $\tau(fg) = f(o)\,\tau(g) + g(o)\,\tau(f)$ for all $f, g \in A^0 \simeq R[[X_1, \ldots, X_n]]$, or, equivalently, is simply an R-linear map of $I^0/(I^0)^2$ into L, where $I^0 = (X_1, \ldots, X_n)$ is the augmentation ideal in A^0 (namely the map induced by the restriction of τ to I^0). Thus, $t_G(L)$ is a vector space of dimension $n = \dim G$ over L. The logarithm map: $\log: G(S) \to t_G(L)$ is defined as follows:

$$(\log x)(f) = \lim_{i \to \infty} \left(\frac{f(p^i x) - f(o)}{p^i} \right),$$

for $x \in G(S)$ and $f \in A^0$; note that for large i we will have $p^i x \in G^0(S)$, because $G^{et}(S)$ is a torsion group. Alternatively, we can identify $I^0/(I^0)^2$

with the space of invariant differential forms ω on Γ, and define, for $x \in G^0(S)$ which is all that really matters,

$$(\log x)(\omega) = \Omega(x),$$

where $\Omega(X) \in K[[X_1, \ldots, X_n]]$ is such that $\Omega(0)=0$ and $d\Omega=\omega$, see SERRE [*12*]. Using either of these definitions of log, one proves easily that it is a $\mathbf{Z}_p$-homomorphism, and a local isomorphism. More precisely, that if $c^{p-1}<|p|$, the logarithm gives an isomorphism between the group of points $x=(x_i)$ in $G^0(S)$ such that $|x_i| \leqslant c$ for all i and the group of points $\tau \in t_G(L)$ such that $|\tau(X_i)| \leqslant c$ for all i. From these facts it follows that the kernel of log is the torsion subgroup of $G(S)$, and that its cokernel is a torsion group. Thus, the log induces an *isomorphism*

$$G(S) \underset{\mathbf{Z}_p}{\otimes} \mathbf{Q}_p \xrightarrow{\sim} t_G(L).$$

It also follows that log $G(S)$ is contained in a finitely generated S-submodule of $t_G(L)$ if the valuation on S is discrete, whereas $\log G(S)=t_G(L)$ if L is algebraically closed.

Examples. 1.) If $G=\mathbf{G}_m(p)$, then $G(S)$ is the group of units congruent to 1 in S, $t_G(L)$ is L, and the logarithm is the ordinary p-adic logarithm.

2.) If X is an abelian scheme over R, and $G=X(p)$, then we can identify $G(S)$ with the subgroup of $X(S)$ consisting of the points x whose reduction mod the maximal ideal of S is of finite p-power order, $G^0(S)$ being identified with the kernel of the reduction map. The logarithm is then the map which has been studied by LUTZ (for elliptic curves) and by MATTUCK in general.

The Galois modules Φ and T. Let $\bar{K}$ be the algebraic closure of K, and $\mathscr{G}=\mathrm{Gal}(\bar{K}/K)$. Put

$$\Phi(G) = \lim_{\substack{\longleftarrow \\ \nu}} G_\nu(\bar{K}), \quad \text{with respect to the maps} \quad i_\nu: G_\nu \hookrightarrow G_{\nu+1}.$$

$$T(G) = \lim_{\substack{\longleftarrow \\ \nu}} G_\nu(\bar{K}), \quad \text{with respect to the maps} \quad j_\nu: G_{\nu+1} \to G_\nu.$$

Since $\mathrm{char}(K)=0$, the $G_\nu \otimes_R K$ are étale, and it follows from the definition of p-divisible groups that $\Phi(G)$ and $T(G)$ are $\mathbf{Z}_p$-modules isomorphic respectively, to $(\mathbf{Q}_p/\mathbf{Z}_p)^h$ and to $\mathbf{Z}_p^h$, where $h=\mathrm{height}(G)$, on which $\mathscr{G}$ acts continuously. We have canonical isomorphisms:

$$\Phi(G) \simeq T(G) \underset{\mathbf{Z}_p}{\otimes} (\mathbf{Q}_p/\mathbf{Z}_p) \quad \text{and} \quad T(G) \simeq \mathrm{Hom}_{\mathbf{Z}_p}(\mathbf{Q}_p/\mathbf{Z}_p, \Phi(G)),$$

so knowledge of $\Phi(G)$ is equivalent to that of $T(G)$, and knowledge of either is equivalent to knowledge of the *general fiber* $G \otimes_R K$ of the p-divisible group G.

In the notation of the preceding paragraphs, the map $G_v(S) \to G_v(L)$ is bijective, so that, $\Phi(G)$ is the torsion subgroup of $G(S)$ if L is the completion of $\bar{K}$.

Examples. 1.) If $G = \mathbf{G}_m(p)$, then $\Phi(G)$ is the group of roots of unity of prime power order in $\bar{K}$.

2.) If X is an abelian scheme of dimension n over R, and $G = X(p)$, then $T(G) = T_p(X)$, the p-adic representation space of rank $h = 2n$ introduced by WEIL.

§ 3. The completion of the algebraic closure of K

R denotes a discrete valuation ring, which is complete, of characteristic 0. Its residue field k is assumed to be perfect of characteristic $p > 0$. Let K denote the quotient field of R, $\bar{K}$ its algebraic closure and C the completion of $\bar{K}$. C is known to be algebraically closed (see [*1*], p. 42). The absolute value on K is canonically extended to C. Let π be a uniformizing element of K. For $x \in C$ we define its order $v(x)$ by

$$|x| = |\pi|^{v(x)},$$

we put $v(p) = e$.

If I is an ideal in some finite extension of K, $v(I)$ is defined in the obvious way. The relative different of a finite extension M/L is denoted by $\mathfrak{d}_{M/L}$.

In the results of this section several constants will appear. We often will denote them indiscriminately by the same letter.

(3.1). Study of certain totally ramified extensions.

Let K_∞ be an infinite Galois extension of K which is totally ramified with group $\mathscr{C} \simeq \mathbf{Z}_p$. Let K_n be the subfield of K_∞ which corresponds to the closed subgroup $\mathscr{C}(n) = p^n \mathbf{Z}_p$. Then K_n/K is cyclic of degree p^n.

Proposition 5. *There is a constant c such that*

$$v(\mathfrak{d}_{K_n/K}) = en + c + p^{-n} a_n,$$

where a_n is bounded.

This follows from standard facts about higher ramification, together with local class field theory.

We use the higher ramification groups $\mathscr{C}^v$, with the "upper numbering", which is compatible with passage to quotients ([*8*], p. 119, or [*2*], p. 81). Suppose that $\mathscr{C}^v = \mathscr{C}(i)$ for $v_i < v \leqslant v_{i+1}$. We then have $v_{n+1} = v_n + e$ for large n (recall that $e = v(p)$). This follows from local class field theory, more precisely from the description of the images of the $\mathscr{C}^v$ under the reciprocity mapping. (See [*8*], p. 235, Cor. 3, for the case of a finite k; see [*9*] for the case of an algebraically closed k, to which the case of arbitrary perfect k can be reduced by passage to the completion of the maximal unramified extension of K; see also the Berkeley Ph. D. thesis of B. F. WYMAN (June 1966).) Put $\mathfrak{d}_n = \mathfrak{d}_{K_n/K}$. A well-known formula ([*8*], p. 109) then gives ($o(\mathscr{G})$ denoting the order of a group $\mathscr{G}$)

$$v(\mathfrak{d}_n) = \int_{-1}^{\infty} (1 - o(\mathrm{Gal}(K_n/K)^v)^{-1})\, dv .$$

Now $\mathrm{Gal}(K_n/K)^v = \mathscr{C}^v \mathscr{C}(n)$, whence

$$\begin{aligned} o(\mathrm{Gal}(K_n/K)^v) &= p^{n-i} \quad \text{if} \quad v_i < v \leqslant v_{i+1}, \quad \text{with} \quad i \leqslant n, \\ &= 1 \quad \text{otherwise}. \end{aligned}$$

The assertion then follows easily.

Corollary 1. $v(\mathfrak{d}_{K_{n+1}/K_n}) = e + p^{-n} b_n$, *where* b_n *is bounded.*

Corollary 2. *There is a constant* a *(independent of* n*) such that for* $x \in K_{n+1}$ *we have*

$$|\mathrm{Tr}_{K_{n+1}/K_n}(x)| \leqslant |p|^{1-ap^{-n}} |x| .$$

Let R_n be the ring of integers of K_n and $\mathfrak{m}_n$ its maximal ideal. Let $\mathfrak{d}_{K_{n+1}/K_n} = \mathfrak{m}_{n+1}^d$. Then

$$\mathrm{Tr}_{K_{n+1}/K_n}(\mathfrak{m}_{n+1}^i) = \mathfrak{m}_n^j ,$$

where $j = \left[\dfrac{i+d}{p}\right]$ (see [*8*], p. 91, Lemma 4). This implies the asserted inequality.

Corollary 3. *There is a constant* c *(independent of* n*) such that for* $x \in K_n$ *we have*

$$|\mathrm{Tr}_{K_n/K}(x)| \leqslant |p|^{n-c} |x| .$$

Let σ denote a generator of the group $\mathscr{C}$.

Lemma 2. *There exists a constant* $c > 0$ *(independent of* n*) such that for* $x \in K_{n+1}$ *we have*

$$|x - p^{-1}\, \mathrm{Tr}_{K_{n+1}/K_n}(x)| \leqslant c\, |\sigma^{p^n} x - x| .$$

Proof. Let $\tau=\sigma^{p^n}$. Then

$$px-\mathrm{Tr}_{K_{n+1}/K_n}(x)=px-\sum_{i=0}^{p-1}\tau^i x=$$
$$\sum_{i=0}^{p-1}(1-\tau^i)\,x=\sum_{i=1}^{p-1}(1+\tau+\cdots+\tau^{i-1})(1-\tau)\,x.$$

Hence

$$|px-\mathrm{Tr}_{K_{n+1}/K_n}(x)|\leqslant|(1-\tau)\,x|,$$

and we may take $c=|p|^{-1}$.

Now define a K-linear function t on K_∞ with values in K as follows: For $x\in K_n$, put

$$t(x)=p^{-n}\,\mathrm{Tr}_{K_n/K}(x).$$

This is independent of the choice of n.

Proposition 6. *There exists a constant $d>0$ such that we have for all $x\in K_\infty$,*

$$|x-t(x)|\leqslant d\,|\sigma x-x|.$$

We prove by induction on n an inequality

$$|x-t(x)|\leqslant c_n\,|\sigma x-x|\quad\text{if}\quad x\in K_n, \tag{*}$$

with

$$c_{n+1}=|p|^{-ap^{-n}}c_n,$$

where a is a constant >0. This will imply the assertion. We may take c_1 equal to the c of Lemma 2. Then we have for $x\in K_{n+1}$, assuming (*) to be true,

$$|\mathrm{Tr}_{K_{n+1}/K_n}(x)-pt(x)|\leqslant c_n\,|\sigma\,\mathrm{Tr}_{K_{n+1}/K_n}(x)-\mathrm{Tr}_{K_{n+1}/K_n}(x)|$$
$$=c_n\,|\mathrm{Tr}_{K_{n+1}/K_n}(\sigma x-x)|\leqslant c_n\,|p|^{1-ap^{-n}}\,|\sigma x-x|,$$

by Cor. 2 to Proposition 5. By Lemma 2 we have then

$$|x-t(x)|\leqslant\mathrm{Max}\big(|x-p^{-1}\,\mathrm{Tr}_{K_{n+1}/K_n}(x)|,\ \ |p|^{-ap^{-n}}c_n\,|\sigma x-x|\big)$$
$$\leqslant\mathrm{Max}\big(c_1,\,|p|^{-ap^{-n}}c_n\big)\,|\sigma x-x|=|p|^{-ap^{-n}}c_n\,|\sigma x-x|,$$

which establishes (*) for $(n+1)$.

Remark. From the proof we see that if we take K_n as a groundfield instead of K we have a corresponding inequality with *the same constant d.*

Next let X be the completion of K_∞. This is a Banach space over K, on which $\mathscr{C}$ acts continuously. The preceding results will enable us to get some information about the $\mathscr{C}$-space X.

By Prop. 6 (or by Cor. 3 of Prop. 5) the linear operator t is continuous on K_∞ and therefore extends to X by continuity. We have $t(X)=K$ because K is complete; let X_0 be the kernel of t on X, a closed subspace of X.

Proposition 7. a) *X is the direct sum of K and X_0.*

b) *The operator $\sigma-1$ annihilates K, and is bijective, with a continuous inverse, on X_0.*

c) *Let λ be a unit in K which is $\equiv 1 \pmod{\pi}$ and which is not a root of unity. Then $\sigma-\lambda$ is bijective, with a continuous inverse, on X.*

Proof. (a) Since t is *idempotent*, X is the direct sum of its range and its kernel.

(b) For each n, let $K_{n,0}=K_n \cap X_0$ be the subspace of K_n consisting of the elements whose trace to K is 0, and let $K_{\infty,0}=\bigcup_{n=0}^{\infty} K_{n,0}$. We have $K_n=K\oplus K_{n,0}$, direct sum, for $0\leqslant n\leqslant\infty$, and X_0 is the closure of $K_{\infty,0}$ in X. The operator $\sigma-1$ is injective, hence bijective, on each of the finite dimensional spaces $K_{n,0}$, for $n<\infty$, and is therefore bijective on their union $K_{\infty,0}$; let ϱ be its inverse. By Prop. 6 we have $|\varrho y|\leqslant d|y|$ for each $y=(\sigma-1)x$ in $K_{\infty,0}$. Hence ϱ extends by continuity to X_0, and this extension is a continuous inverse for $\sigma-1$ on X_0.

(c) Since $\sigma-\lambda$ is obviously bijective on K for $\lambda\neq 1$, we can, by (a), restrict our attention to its action on X_0. As operators on X_0 we have

$$\varrho(\sigma-\lambda)=\varrho((\sigma-1)-(\lambda-1))=1-(\lambda-1)\varrho. \qquad (*)$$

If $|\lambda-1|\,d<1$ with d as in (b), we have $|(\lambda-1)\,\varrho(y)|<|y|$ for all $y\in X_0$ and consequently $1-(\lambda-1)\,\varrho$ is an automorphism of X_0, its inverse being given by a geometric series, and consequently, by (*), $\sigma-\lambda$ has a continuous inverse on X_0. If $|\lambda-1|\,d\geqslant 1$, we replace σ by σ^{p^n}, and λ by λ^{p^n}, where n is so large that $|\lambda^{p^n}-1|\,d<1$. We then replace K by K_n. Taking into account the remark following Prop. 6, we find from what precedes that $\sigma^{p^n}-\lambda^{p^n}$ has a bounded inverse on X, whence also $\sigma-\lambda$.

We can define, in the obvious way, cohomology groups $H^i(\mathscr{C}, X)$ based on continuous cochains for the canonical topologies on $\mathscr{C}$ and X. If χ is a continuous character of $\mathscr{C}$ into the group of units of K, then we denote by $X(\chi)$ the space X with the "twisted" action

$${}^s x=\chi(s)\,(sx),$$

for $s\in\mathscr{C}$, $x\in X$.

We then have

Proposition 8. (a) $H^0(\mathscr{C}, X)=K$, *and* $H^1(\mathscr{C}, X)$ *is a one-dimensional vector space over* K.

(b) *If* $\chi(\mathscr{C})$ *is infinite, then* $H^0(\mathscr{C}, X(\chi))$ *and* $H^1(\mathscr{C}, X(\chi))$ *are* 0.

Proof. If Y is a closed subspace of X stable under $\mathscr{C}$, then $H^0(\mathscr{C}, Y(\chi))$ is the kernel of $\sigma-\lambda$ on Y, and, since a 1-cocycle on $\mathscr{C}$ is determined by its value at σ, $H^1(\mathscr{C}, Y(\chi))$ is a subgroup of the cokernel of $\sigma-\lambda$ on Y. In particular, both cohomology groups vanish if $\sigma-\lambda$ is bijective on Y. Hence by part (b) of Prop. 7 we see that $H^0(\mathscr{C}, X_0)$ and $H^1(\mathscr{C}, X_0)$ both vanish, and consequently (a) follows from part (a) of Prop. 7. Similarly, (b) follows from part (c) of Prop. 7, because if $\chi(\mathscr{C})$ is infinite, then $\lambda=\chi(\sigma)$ is not a root of unity.

(3.2). Finite extensions of K_∞

We keep the notations of (3.1). Let L denote a finite extension of K_∞. Denote by R_L its ring of integers and by $\mathfrak{m}_L$ its maximal ideal. R_∞ and $\mathfrak{m}_\infty$ have the same meaning for K_∞. Let $\mathscr{H}=\mathrm{Gal}(\bar{K}/K_\infty)$.

Proposition 9. *We have* $\mathrm{Tr}_{L/K_\infty}(R_L)\supset\mathfrak{m}_\infty$.

Replacing K by one of the K_n we may assume that there is a finite extension L_0 of K, linearly disjoint from K_∞ over K, such that $L=L_0K$ (see [8], p. 97, Lemma 6).

We may also suppose that L_0/K is a Galois extension. Put $L_n=L_0K_n$. Then

$$v(\mathfrak{d}_{L_n/K_n})=\int_{-1}^{\infty}\left(o(\mathrm{Gal}(K_n/K)^v)^{-1}-o(\mathrm{Gal}(L_n/K)^v)^{-1}\right)dv\,.$$

Let $\mathrm{Gal}(L/K)^v\subset\mathscr{C}$, i.e., $\mathrm{Gal}(L_0/K)^v=(1)$, for $v\leqslant h$. It follows that

$$v(\mathfrak{d}_{L_n/K_n})\leqslant\int_{-1}^{h}o(\mathrm{Gal}(K_n/K)^v)^{-1}\,dv\,.$$

The argument used in the proof of Prop. 5 now shows that this tends to zero with n (the order of magnitude is p^{-n}). The assertion then follows from familiar results ([8], p. 91, Lemma 4).

Corollary 1. *Let* L/K_∞ *be finite Galois with group* G. *Let* f *be an* r*-cochain of* G *with coefficients in* L, *with* $r\geqslant 0$, *and let* $c>1$. *Then there exists an* $(r-1)$*-cochain* g *of* G *in* L *such that*

$$|f-\delta g|\leqslant c\,|\delta f|\,,\quad\text{and}\quad |g|\leqslant c\,|f|\,.$$

Here $|f|$ denotes the maximum of the absolute values of the coefficients of f, and by a (-1)-cochain we mean an element $y \in L$. The coboundary δy of such a y is the 0-cochain Tr_{L/K_∞}.

Proof. By Prop. 9 there exists a (-1)-cochain $y \in L$ such that $|y| \leq 1$ and $|\delta y| > c^{-1}$. Define an $(r-1)$-cochain $y \cup f$ by the formulas

$$y \cup f = yf, \quad \text{if} \quad r = 0$$

$$(y \cup f)(s_1, \dots, s_{r-1}) = (-1)^r \sum_{s_r \in G} s_1 s_2 \cdots s_r y \cdot f(s_1, s_2, \dots, s_r), \quad \text{if} \quad r > 0.$$

The identity

$$(\delta y) f - \delta(y \cup f) = y \cup (\delta f)$$

is easily checked; on dividing by the element $x = \delta y = \mathrm{Tr}_{L/K_\infty} y \in K$ we find

$$f - \delta g = x^{-1}(y \cup \delta f), \quad \text{with} \quad g = x^{-1}(y \cup f).$$

Since $|x^{-1}| < c$, and $|y| \leq 1$, the corollary follows.

Corollary 2. *The corollary 1 still holds true if we replace L by $\bar{K}$, G by $\mathscr{H}$, and consider cochains which are continuous from the Krull topology in G to the discrete topology in $\bar{K}$, provided that, for $r=0$, the conclusion is replaced by: there exists an element $x \in K_\infty$ such that $|f - x| \leq c|\delta f|$.*

This follows from Cor. 1, because a continuous cochain in $\mathscr{H}$ with values in $\bar{K}$ comes by inflation from some finite Galois L/K_∞ (cf. Serre, Cohomologie Galoisienne, Prop. 8.)

Recall that we denote by C the completion of $\bar{K}$. By $H^r(\mathscr{H}, C)$ we mean the cohomology groups constructed with standard cochains which are continuous from the Krull topology in $\mathscr{H}$ to the *valuation topology* in C.

Proposition 10. *We have $H^0(\mathscr{H}, C) = X$, and $H^r(\mathscr{H}, C) = 0$, for $r > 0$.*

Proof. This will follow immediately from Corollary 2, once we show that, for every continuous cochain f on $\mathscr{H}$ with values in C, there is a sequence of cochains f_ν on $\mathscr{H}$ with values in $\bar{K}$ as in Cor. 2, such that $|f - f_\nu| \to 0$. To construct such f_ν, let D be the ring of integers in C. Then $C = \bar{K} + \pi^\nu D$ for each ν, and there exist maps $\varphi_\nu : C/\pi^\nu D \to \bar{K}$ such that $\psi_\nu \varphi_\nu = \mathrm{id.}$, where $\psi_\nu : C \to C/\pi^\nu D$ is the canonical projection. The φ_ν are automatically continuous because $C/\pi^\nu D$ is discrete. Put $f_\nu = \varphi_\nu \psi_\nu f$. Then $\psi_\nu f_\nu = f_\nu$ implies $|f_\nu - f| \leq |\pi|^\nu$.

(3.3). The action of $\mathscr{G}$ on C

Let $\mathscr{G}$ denote the Galois group of $\bar{K}/K$. Then $\mathscr{G}$ operates on C by continuity and we can consider the continuous cochain cohomology groups $H^r(\mathscr{G}, C)$.

Theorem 1. *We have $H^0(\mathcal{G}, C)=K$, and $H^1(\mathcal{G}, C)$ is a one-dimensional vector space over K.*

Proof. Let K_∞/K be as in 3.1; for example, we can take for K_∞ a suitable subfield of the field generated over K by all p^n-th roots of 1, all n. Then we have an isomorphism

$$H^0(\mathcal{G}, C) \simeq H^0(\mathcal{G}/\mathcal{H}, H^0(\mathcal{H}, C))$$

and an exact inflation-restriction sequence

$$0 \to H^1(\mathcal{G}/\mathcal{H}, H^0(\mathcal{H}, C)) \to H^1(\mathcal{G}, C) \to H^1(\mathcal{H}, C),$$

from which our theorem follows, using Prop. 10 and Prop. 8(a).

Remarks. 1. The one-dimensionality of $H^1(\mathcal{G}, C)$ was not known to me for arbitrary K at the time of the conference. It was proved by T. SPRINGER when he wrote up the first draft of these notes.

2. Let $\bar{R}$ denote the integral closure of R in $\bar{K}$, with the *discrete* topology. The methods we have used here yield very easily the fact that $H^1(\mathcal{G}, \bar{R})$ is killed by some power of p (the power depending perhaps on K), and this fact in turn implies easily the first part of Theorem 1, i.e., $C^{\mathcal{G}}=K$. Meanwhile, Shankar SEN has shown that $H^1(\mathcal{G}, \bar{R})$ is killed by p for p odd, and by 4 if $p=2$. From this result of SEN it follows easily that for every closed subgroup $\mathcal{G}_1$ of $\mathcal{G}$ we have $C^{\mathcal{G}_1}=\hat{K}_1$ where $K_1=\bar{K}^{\mathcal{G}_1}$, and where "hat" denotes completion.

3. Recently, James Ax has given a short proof of this last result, by a direct method which avoids the use of higher ramification theory and of the intermediate field K_∞.

Now let $\chi:\mathcal{G}\to K^*$ be a continuous homomorphism (note that the values of χ are units in K^* because $\mathcal{G}$ is compact), and let $C(\chi)$ denote C with the twisted action ${}^s x=\chi(s)\, sx$. Let K_∞ denote the extension of K determined by Ker χ.

Theorem 2. *Suppose that there is a finite extension K_0 of K contained in K_∞ such that K_∞/K_0 is totally ramified and $\mathrm{Gal}(K_\infty/K_0)\simeq \mathbf{Z}_p$. Then $H^0(\mathcal{G}, C(\chi))=0$ and $H^1(\mathcal{G}, C(\chi))=0$.*

Proof. It is easy to reduce the statement to the case $K=K_0$, and in that case the result follows if we apply Prop. 10 and Prop. 8(b) as in the proof of Theorem 1.

§ 4. Theorems on p-divisible groups

We continue now the discussion of (2.4). Let G be a p-divisible group over our complete discrete valuation ring R of mixed characteristic and let K, C, and D be as in § 3. Let G' be the dual of G. By Cartier duality we have for each ν

$$G'_\nu(D) \xrightarrow{\sim} \mathrm{Hom}_D(G_\nu \underset{R}{\otimes} D, \mathbf{G}_m).$$

Passing to the projective limit as $\nu \to \infty$ we obtain an isomorphism

$$T(G') \xrightarrow{\sim} \mathrm{Hom}_D(G \underset{R}{\hat{\otimes}} D, \mathbf{G}_m(p))$$

where $\mathbf{G}_m(\mathrm{p})$ is the p-divisible group attached to $\mathbf{G}_m$, viewed over D This isomorphism gives us pairings

$$T(G') \times G(D) \to (\mathbf{G}_m(p))(D) \simeq U$$

and

$$T(G') \times t_G(C) \to t_{\mathbf{G}_m(p)}(C) \simeq C,$$

where U denotes the group of units congruent to 1 in D. These pairings are compatible with the logarithm map $L: G(D) \to t_G(C)$ and the ordinary p-adic logarithm $U \to C$. The kernel of these logs is the torsion subgroup of their domain, and they are surjective because, C being algebraically closed, $G(D)$ and U are divisible. Thus we get an exact commutative diagram

$$\begin{array}{ccccccccc} 0 & \longrightarrow & \Phi(G) & \longrightarrow & G(D) & \xrightarrow{L} & t_G(C) & \longrightarrow & 0 \\ & & \downarrow{\scriptstyle \alpha_0} & & \downarrow{\scriptstyle \alpha} & & \downarrow{\scriptstyle d\alpha} & & \\ 0 & \to & \mathrm{Hom}(T', U_{\mathrm{tors}}) & \to & \mathrm{Hom}(T', U) & \to & \mathrm{Hom}(T', C) & \to & 0, \end{array} \qquad (*)$$

where $T' = T(G')$ is free of rank h over $\mathbf{Z}_p$, the Homs in the bottom row are $\mathbf{Z}_p$-homs, and where $U_{\mathrm{tors}} \simeq \Phi(\mathbf{G}_m(p))$ is the group of roots of unity in U. The vertical arrows are $\mathscr{G}$-homomorphisms, $\mathscr{G}$ acting on a homomorphism f by the rule $(sf)(x) = s(f(s^{-1}x))$.

Proposition 11. *α_0 is bijective and α and $d\alpha$ are injective.*

The proof is in a series of steps.

Step 1. *α_0 is bijective.* Indeed, since K is of characteristic 0, Cartier duality gives a perfect duality of finite $\mathscr{G}$-modules

$$G_\nu(C) \times G'_\nu(C) \to U_{\mathrm{tors}}$$

for each ν. The result follows on passage to the limit as $\nu \to \infty$, inductively with the G_ν and projectively with the G'.

Incidentally, if we pass to the limit projectively with both, we find a $\mathcal{G}$-isomorphism

$$T(G) \simeq \mathrm{Hom}(T(G'), H),$$

where $H = T(\mathbf{G}_m(p)) = \varprojlim_\nu$ (group of p^ν-th roots of unity).

Step 2. Ker α *and* Coker α *are vector spaces over* $\mathbf{Q}_p$. Applying the snake-lemma to diagram (*) and using Step 1, we find that α and $d\alpha$ have isomorphic kernels and cokernels. Since $d\alpha$ is C-linear, the result follows.

Step 3. *We have* $G(R) = G(D)^{\mathcal{G}}$ *and* $t_G(K) = t_G(C)^{\mathcal{G}}$. This follows from Theorem 1, i.e., from the fact that $K = C^{\mathcal{G}}$, which of course implies that $R = D^{\mathcal{G}}$.

Step 4. α *is injective on* $G(R)$. Indeed, the kernel of the restriction of α to $G(R)$ is $(\mathrm{Ker}\,\alpha)^{\mathcal{G}}$ by step 3, and is therefore uniquely divisible by p, by step 2. If G is connected it follows that $\mathrm{Ker}\,\alpha \cap G(R) = 0$, because in that case, viewing G as a formal Lie group we see that $\bigcap p^\nu G(R) = 0$, because the valuation on R is *discrete* (if x is a point of $G(R)$, all of whose coordinates are $\equiv 0 \pmod{\pi^i}$ then the coordinates of px are $\equiv 0 \pmod{\pi^{i+1}}$). In the general case we see then that $(\mathrm{Ker}\,\alpha) \cap G^0(R) = 0$, where G^0 is the connected component of G (use the functorality of (*) with respect to $G^0 \to G$, and the fact that $T(G') \to T((G^0)')$ is surjective). Since $\mathrm{Ker}\,\alpha$ is torsion-free and $G(R)/G^0(R)$ is a torsion group, it follows that $\mathrm{Ker}\,\alpha \cap G(R) = 0$ as claimed.

Step 5. *The map* $d\alpha$ *is injective on* $t_G(K)$. From steps 1 and 4 we conclude that $d\alpha$ is injective on $L(G(R))$; but that group spans $t_G(K)$ over $\mathbf{Q}_p$.

Step 6. *The map* $d\alpha$ *is injective.* The arrow $d\alpha$ can be factored as follows

$$t_G(C) \simeq t_G(K) \underset{K}{\otimes} C \to \mathrm{Hom}_{\mathcal{G}}(T', C) \underset{K}{\otimes} C \to \mathrm{Hom}(T', C).$$

The left-hand arrow is injective by step 5. The right-hand one is injective by

Step 7. *Let* W *be a vector space over* C *on which* $\mathcal{G}$ *operates semi-linearly* (i.e., $s(cw) = s(c)\,s(w)$, for $s \in \mathcal{G}$, $c \in C$, $w \in W$). *Then the C-linear map*

$$W^{\mathcal{G}} \underset{K}{\otimes} C \to W$$

is injective. In down-to-earth terms, this statement means that if a family of elements $w_i \in W^{\mathcal{G}}$ is independent over K, then the family is independent over C. It can be proved by looking at a "shortest" hypothetical dependence relation $\sum c_i w_i = 0$, with $c_i = 1$ for some i, applying elements $s \in \mathcal{G}$ to it and using Theorem 1, i.e., the fact that K is the fixed field of the group of automorphisms $\mathcal{G}$ of the field C. See SERRE [10], Prop. 4, for a more general statement.

Proposition 11 now follows from Steps 1 and 7 and the snake-lemma.

Theorem 3. *The maps*

$$G(R) \xrightarrow{\alpha_R} \operatorname{Hom}_{\mathcal{G}}(T(G'), U)$$

and

$$t_G(K) \xrightarrow{d\alpha_R} \operatorname{Hom}_{\mathcal{G}}(T(G'), C)$$

induced by α and $d\alpha$ are bijective.

Proposition 11 implies the injectivity of these maps, and also, via step 3 above, that we have injections

$$\operatorname{Coker}\alpha_R \hookrightarrow (\operatorname{Coker}\alpha)^{\mathcal{G}} \quad \operatorname{Coker}(d\alpha_R) \hookrightarrow (\operatorname{Coker} d\alpha)^{\mathcal{G}}.$$

Since $\operatorname{Coker}\alpha \to \operatorname{Coker} d\alpha$ is bijective, it follows that the map $\operatorname{Coker}\alpha_R \to \operatorname{Coker} d\alpha_R$ is injective, so we are reduced to proving that $d\alpha_R$ is surjective. Since $d\alpha_R$ is K-linear and injective, this is a question of dimensions. Let

$$W' = \operatorname{Hom}(T(G'), C) \quad \text{and} \quad W = \operatorname{Hom}(T(G), C),$$

spaces of dimension $h = \operatorname{ht}(G)$ over C on which $\mathcal{G}$ operates semilinearly. Put

$$d' = \dim_K (W')^{\mathcal{G}} \quad d = \dim_K W^{\mathcal{G}}$$
$$n = \dim G = \dim_K t_G(K) \quad n' = \dim G' = \dim_K t_{G'}(K).$$

By the injectivity of $d\alpha_R$ we already know $n \leqslant d'$ and $n' \leqslant d$, and we wish to show that equality holds. Since $n + n' = h$, it will suffice to show that $d + d' \leqslant h$. This we do as follows.

Since $T(G) \simeq \operatorname{Hom}(T(G'), H)$ (see step 1 of proof of Prop. 11), we have $W' = T(G) \otimes \operatorname{Hom}(H, C)$, so that there is a canonical non-degenerate $\mathcal{G}$-pairing

$$W \times W' \to Y,$$

where $Y = \operatorname{Hom}(H, C)$. This space Y is isomorphic to $C(\chi^{-1})$, where $\chi: \mathcal{G} \to \mathbf{Z}_p^*$ is the character such that $sz = z^{\chi(s)}$ for all roots of unity z of p-power order. Therefore, by Theorem 2, $Y^{\mathcal{G}} = H^0(\mathcal{G}, Y) = 0$, and also $H^1(\mathcal{G}, Y) = 0$. Since the spaces $W^{\mathcal{G}}$ and $(W')^{\mathcal{G}}$ are paired into $Y^{\mathcal{G}}$, it follows that $W^{\mathcal{G}}C$ and $(W')^{\mathcal{G}}C$ are orthogonal C-subspaces of W and W'. Their dimensions are d and d' (step 7 of the proof of Prop. 11). Hence $d + d' \leqslant h = \dim_C W$, as required.

Corollary 1. *The $\mathcal{G}$-module $T(G)$ determines the dimension n of $\mathcal{G}$.*

Indeed, $T(G)$ determines $T(G')$ by duality, and $n=\dim_K(t_G(K))= =\dim_K(\mathrm{Hom}_{\mathcal{G}}(T(G'), C))$ by Theorem 3.

Corollary 2. *There is a canonical isomorphism of $\mathcal{G}$-modules*

$$\mathrm{Hom}(T(G), C) \simeq t_{G'}(C) \oplus t_G^*(C) \underset{C}{\otimes} \mathrm{Hom}(H, C),$$

where t_G^* is the cotangent space of G at the origin.

The proof of Theorem 3 above shows that $d\alpha'$ and $d\alpha$ map $t_{G'}(C)$ and $t_G(C)$ injectively onto subspaces of W and W' which are orthogonal complements with respect to the pairing to Y. Thus we have an exact sequence

$$0 \to t_{G'}(C) \overset{d\alpha'}{\to} W \to \mathrm{Hom}_C(t_G(C), Y) = t_G^*(C) \underset{C}{\otimes} Y \to 0,$$

and to prove the corollary, we must show that this sequence has a unique splitting compatible with the action of $\mathcal{G}$. The sequence has the form

$$0 \to C^{n'} \to W \to C(\chi^{-1})^n \to 0,$$

where χ is the character in the proof of Theorem 3. The existence of a splitting follows from $H^1(\mathcal{G}, C(\chi))=0$; and its unicity from $H^0(\mathcal{G}, C(\chi)) =0$ (cf. Theorem 2).

Remark: In case $G=A(p)$, where A is an abelian scheme over R, Corollary 2 can be rewritten as

$$H^1(A_C, \mathbf{Q}_p) \otimes C \simeq H^1(A_C, \Omega^0_{A_C}) \oplus H^0(A_C, \Omega^1_{A_C}) \otimes \mathrm{Hom}(H, C),$$

where $A_C = A \otimes_R C$, and where on the left we have the étale cohomology of A_C with coefficients in $\mathbf{Q}_p$. One can ask whether a similar Hodge-like decomposition exists for the étale cohomology with values in C in all dimensions, for a scheme X_C coming from a scheme X projective and smooth over R, or perhaps even over K, or for suitable "rigid analytic" spaces

(4.2.). We can now prove the main result

Theorem 4. *Let R be an integrally closed, noetherian, integral domain, whose field of fractions K is of characteristic 0. Let G and H be p-divisible groups over R. A homomorphism $f: G\otimes_R K \to H \otimes_R K$ of the general fibers extends uniquely to a homomorphism $G \to H$.*

Corollary 1. *The map* $\mathrm{Hom}\,(G,\,H) \to \mathrm{Hom}_{\mathscr{G}}(T(G),\,T(H))$ *is bijective, where* $\mathscr{G} = \mathrm{Gal}\,(\bar{K}/K)$.

Corollary 2. *If* $g: G \to H$ *is a homomorphism such that its restriction* $G \otimes_R K \to H \otimes_R K$ *is an isomorphism, then g is an isomorphism.*

Since $R = \cap_P R_P$, where P runs over the minimal non-zero primes of R, and since each R_P is a discrete valuation ring, we are immediately reduced to the case R is a discrete valuation ring. There exists an extension R' of R which is a complete discrete valuation ring with algebraically closed residue field and such that $R = R' \cap K$; hence we may assume R is complete with algebraically closed residue field, k. If $\mathrm{char}\,k \neq p$, then G is étale and the theorem is obvious. Thus we are reduced to the case of an R of the type considered in the preceding paragraphs, which we assume from now on.

We first prove Corollary 2. Let $G = (G_\nu)$ and $H = (H_\nu)$, and let A_ν (resp. B_ν) denote the affine algebra of G_ν (resp. H_ν). We are given a coherent system of homomorphisms $u_\nu: B_\nu \to A_\nu$, of which we know that their extensions $u_\nu \otimes 1: B_\nu \otimes_R K \to A_\nu \otimes_R K$ are isomorphisms. Since B_ν is free over R, it follows that u_ν is injective for all ν. To prove surjectivity, we look at the discriminants of the R-algebras A_ν and B_ν. By Prop. 2, these discriminants are non-zero, and are determined by the heights of G and H and their dimensions. But the height and dimension of a p-divisible group over R are determined by its general fiber, the height trivially, and the dimension by Cor. 1 of Theorem 3, since the general fiber of G determines the $\mathscr{G}$-module $T(G)$. Hence the discriminants of A_ν and B_ν are equal and non-zero, and it follows that u_ν is bijective. This proves Corollary 2.

To derive the theorem from the corollary, we will use

Proposition 12. *Suppose F is a p-divisible group over R, and M a $\mathscr{G}$-submodule of $T(F)$ such that M is a $\mathbf{Z}_p$-direct summand. Then there exists a p-divisible group Γ over R and a homomorphism $\varphi: \Gamma \to F$ such that φ induces an isomorphism $T(\Gamma) \xrightarrow{\sim} M$.*

Granting this Proposition we prove the theorem, letting $F = G \times H$, and letting M be the graph of the homomorphism $T(G) \to T(H)$ which corresponds to the given homomorphism $f: G \otimes_R K \to H \otimes_R K$. By Prop. 12 we get a p-divisible group Γ over R and a homomorphism $\varphi: \Gamma \to G \times H$ such that the composition $\mathrm{pr}_1 \cdot \varphi: \Gamma \to G$ induces an isomorphism $T(\Gamma) \to T(G)$, hence an isomorphism on the general fibers. By Cor. 2, it follows that $\mathrm{pr}_1 \cdot \varphi$ is an isomorphism. Thus $\mathrm{pr}_2 \cdot \varphi \cdot (\mathrm{pr}_1 \cdot \varphi)^{-1}: G \to H$ is a homomorphism extending f. The unicity of such an extension is obvious, and this concludes the proof of Theorem 4.

Proof of Prop. 12. The submodule $M \subset T(F)$ corresponds to a closed

p-divisible subgroup $E_* \subset F \otimes_R K$. Let E be the "closure of E_* in F". By this we mean the following: Let B_ν be the affine R-algebra of F_ν, let $A_{*\nu}$ be the affine K-algebra of $E_{*\nu}$, and let $u_\nu: B_\nu \otimes_R K \to A_{*\nu}$ correspond to the inclusion $E_{*\nu} \subsetneq F_\nu \times_R K$. Put $A_\nu = u_\nu(B_\nu)$, and put $E_\nu = \operatorname{Spec} A_\nu$. Then E_ν is a closed subgroup of F_ν for each ν, and the inclusions $F_\nu \to F_{\nu+1}$ induce inclusions $E_\nu \to E_{\nu+1}$; we put $E = \varinjlim(E_\nu)$. Although E itself may not be p-divisible (see example below), nevertheless $E \times_R K = E_*$ is p-divisible, and it follows that E_{i+1}/E_i is killed by p, hence that p induces homomorphisms

$$E_{i+\nu+1}/E_{i+1} \to E_{i+\nu}/E_i$$

which are isomorphisms on the general fiber. Let D_i be the affine algebra of E_{i+1}/E_i. Then all $D_i \otimes_R K$ can be identified, and the D_i constitute an increasing sequence of orders in a finite separable K-algebra. Hence there is an i_0 such that $D_i = D_{i+1}$ for $i \geqslant i_0$. Put $\Gamma_\nu = E_{i_0+\nu}/E_{i_0}$. Then p^{i_0} induces a coherent collection of homomorphisms $\Gamma_\nu \to E_\nu/E_0 = E_\nu$, which are isomorphisms at the general fiber, and we will therefore be done if we show that $\Gamma = \bigcup \Gamma_\nu$ is p-divisible. For this, we factor the homomorphism p^ν in $\Gamma_{\nu+1}$ as follows

$$\begin{array}{ccc} \Gamma_{\nu+1} = E_{i_0+\nu+1}/E_{i_0} & \xrightarrow{p^\nu} & E_{i_0+\nu+1}/E_{i_0} = \Gamma_{\nu+1} \\ \downarrow \alpha & & \uparrow \gamma \\ E_{i_0+\nu+1}/E_{i_0+\nu} & \xrightarrow{\beta} & E_{i_0+1}/E_{i_0}, \end{array}$$

where α is the canonical projection, γ the canonical inclusion, and where β is induced by p^ν, and is therefore an isomorphism by our choice of i_0. It follows that the kernel of p^ν in $\Gamma_{\nu+1}$ is the same as $\operatorname{Ker}\alpha = \Gamma_\nu$, so Γ is p-divisible as claimed.

The following example, due to SERRE, shows that the map φ in Prop. 12 need not be a closed immersion. Let X be an elliptic curve over R whose reduction $\tilde{X}$ has Hasse invariant $\neq 0$, and suppose the points of order p on X are rational. Then there exist two such points, say x and y, which are independent, but such that $\tilde{x} = \tilde{y}$ is of order p, and the sequence

$$0 \to X \xrightarrow{\varphi} (X/\mathbf{F}_p x) \times (X/\mathbf{F}_p y) \to \operatorname{Coker} \varphi \to 0$$

is then exact over K, but φ is not injective over R, because $\varphi\tilde{x} = 0$. Passing to the associated p-divisible groups, one gets the desired example.

References

[1] ARTIN, E.: Algebraic numbers and algebraic functions. Lecture notes. New York 1950/51.

[*2*] Artin, E. and J. Tate: Class Field Theory. Harvard 1961.
[*3*] Cartier, P.: Colloque sur la theorie des groupes algébriques. Brussels 1962.
[*4*] Demazure, M. and A. Grothendieck: Schémas en groupes. Séminaire I.H.E.S. 1963–64.
[*5*] Grothendieck, A.: Eléments de géométrie algébrique, I.H.E.S. Publications, Paris.
[*6*] Manin, Yu. I.: Theory of formal commutative groups (translation). Russian Math. Surveys, vol. 18, p. 1–83 (1963).
[*7*] Raynaud, M.: Passage au quotient par une relation d'equivalence plate. This volume p. 78–85.
[*8*] Serre, J.-P.: Corps locaux. Hermann, Paris, 1962.
[*9*] Serre, J.-P.: Sur les corps locaux à corps de restes algébriquement clos. Bull. Soc. Math. de France 89, p. 105–154 (1961).
[*10*] Serre, J.-P.: Sur les groupes de galois attachés aux groupes *p*-divisibles. This volume p. 118–131.
[*11*] Serre, J.-P.: Groupes *p*-divisibles (d'après J. Tate). Séminaire Bourbaki, N^0. **318**, (Nov. 1966).
[*12*] Serre, J.-P.: Lie algebras and Lie groups. New York: Benjamin 1965.

A Duality Theorem in the Etale Cohomology of Schemes

J. L. VERDIER

We shall present in this exposé a duality theorem which has been proved by A. GROTHENDIECK. The formulation of this theorem is the same as those of the other duality theorems which can be found in nature: Duality for coherent sheaves [H. S.], duality in the cohomology of pro-finite groups, Poincaré's duality for topological varieties, ...

To get a duality theorem, we need a theory of cohomology with compact support (§ 2). Then the duality is defined by the Gysin's morphism (or trace morphism) (§ 3). In § 1, we shall recall the base change theorem for the étale cohomology which is the main instrument in this question.

§ 1. The base change theorem in the étale cohomology of schemes

Let us consider a cartesian square of preschemes:

$$\begin{array}{ccc} X & \xleftarrow{g'} & X' \\ {\scriptstyle f}\downarrow & & \downarrow{\scriptstyle f'} \\ S & \xleftarrow{g} & S' \end{array}$$

and let F be a torsion sheaf on X for the étale topology. Let us suppose (to simplify) that the prescheme S is locally noetherian. The obvious natural transformation of functors:

$$f_* \to g_* \, f'_* g'^*$$

(lower star = direct image, upper star = inverse image)
yields natural morphisms:

$$g^* R^q f_* (F) \to R^q f'_* g^* (F)$$

1.1 Theorem. (ARTIN-GROTHENDIECK): *The above morphisms are isomorphisms in the two following cases:*

1) *The morphism f is proper.*

2) *The torsion of F is prime to the residual characteristics of S. The morphism g is smooth.*

§ 2. The direct image functor with proper support

Let $f:X\to S$ be a quasi-projective morphism of preschemes where S is locally noetherian. Let $i:X\to X'$ be an S-imersion of X into a prescheme X' projective on S. For any torsion sheaf on X (for the étale topology), we shall denote by $R^q f_!(F)$ the sheaf on S:

$$R^q f_!(F) = R^q f'_*(i_!(F))$$

where $i_!(F)$ is the sheaf on X' obtained by extending F by zero and where $R^q f'_*$ is the q-th derived functor of the direct image by the morphism $f':X'\to S$.

When $S=\operatorname{spec}(\mathbf{C})$ (the field of complex numbers) and X is a non-singular quasi-projective variety, the $R^q f_!$ are isomorphic to the cohomology groups with compact support of the corresponding topological variety. (Comparison theorem).

The sequence $R^q f_!$ $(0\leqslant q)$ is a δ-functor. It can be shown that it is in general *not* a derived functor.

The $R^q f_!$ will be called the δ-functor direct image with proper support. In order to give a sense to this definition we need the

2.1. Proposition. *The δ-functor $R^q f_!$ does not depend on the immersion i into a prescheme projective on S.*

Proof: Let $i:X\to X'$ and $i':X\to X''$ be two S-immersions. We shall only prove that there exists an isomorphism functorial in F:

$$R^q f'_*(i_!(F)) \xrightarrow{\sim} R^q f''_*(i'_!(F))$$

Making use of the fibered product, we can suppose that there exists an S-morphism $g:X'\to X''$ such that the following diagram is commutative:

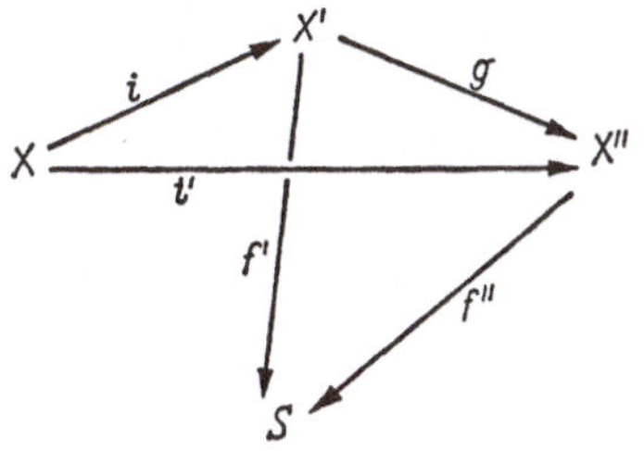

The composition spectral sequence gives:

$$R^p f''_*(R^q g_*(i_! F)) \Rightarrow R^n f'_*(i_! F)$$

To determine the sheaf $R^q g^*(i_!(F))$, we can consider the fibers and apply the base change theorem for a proper morphism. It becomes therefore clear that:

$$R^q g_*(i_!(F)) = 0 \qquad q > 0$$

and that the canonical morphism $i'_!(F) \to g_*(i_!(F))$ is an isomorphism. What remains to be shown is that those various isomorphisms are compatible. This can be done by the same methods.

The properties of the functor $R^q f_!$ are summed up in the following

2.2. Proposition. 1) *The functor $R^q f_!$ commutes with the change of the base.*

1)′ *When f is quasi-finite, $R^q f_! = 0$ $(q \neq 0)$. In particular when f is étale $R^q f_! = 0$ $(q \neq 0)$ and $f_!$ is the functor extension by zero.*

2) *We have a spectral sequence of composition.*

3) *Let $Y \subsetneq X$ be a closed sub-prescheme and U the complementary open sub-prescheme. Let $F_{U!}$ be the sheaf restricted to U and extended by zero on X and F_Y be the direct image on X of the restriction of F on Y.*

We get an unrestricted exact sequence:

$$\cdots \to R^q f_!(F_{U!}) \to R^q f_!(F) \to R^q f_!(F_Y) \to R^{q+1} f_!(F_{U!}) \to \cdots$$

4) *Let $(U_\alpha \to X)$ be a separated étale covering of X. For any simplex $\sigma = (\alpha_1, \alpha_2, \dots, \alpha_p)$ we shall denote by U_σ the prescheme $U_{\alpha_1} \times_S \dots \times_S U_{\alpha_p}$ and by f_{U_σ} the morphism f composed with the canonical morphism $U_\sigma \to X$. Let us denote by F/U_σ the inverse image of the sheaf F on U_σ. For any $q \geqslant 0$ and for any simplicial application $\sigma \to \sigma'$, we get a morphism of sheaves on S:*

$$R^q f_{U_\sigma !}(F/U_\sigma) \to R^q f_{U_{\sigma'} !}(F/U_{\sigma'})$$

which yields a semi-simplicial complex

$$\cdots \overset{\to}{\underset{\to}{\rightrightarrows}} \coprod_{\alpha,\beta} R^q f_{U_\alpha \times_S U_\beta !}(F/U_\alpha \times {}_S U_\beta) \rightrightarrows \coprod_\alpha R^q f_{U_\alpha !}(F/U_\alpha)$$

Let us denote by $H_p(R^q f_{U!}, F)$ the p-th homology sheaf of the above complex.

If the covering is finite or if the dimension of the fibers of the morphism f is bounded, we get a spectral sequence:

$$E_2^{p,q} = H_{-p}(R^q f_{U!}, F) \Rightarrow R^* f_!(F).$$

When the fibers of the morphism f are of dimension $\leqslant d$, the above spectral sequence yields the exact sequence:

$$\coprod_{\alpha,\beta} R^{2d} f_{U_\alpha \times_S U_\beta !}(F/U_\alpha \times {}_S U_\beta) \rightrightarrows \coprod_{\alpha} R^{2d} f_{U_\alpha !}(F/U_\alpha) \to R^{2d} f_!(F) \qquad (2.2.1)$$

Proof. The first three assertions are obvious. Let us prove the fourth one. Let us denote by $F_{U_\sigma !}$ the sheaf restricted to U_σ and extended by zero. The complex of sheaves on $X: C^*(U_\alpha, F)$ deduced from the semi-simplicial complex

$$\overset{\to}{\underset{\to}{\to}} \coprod_{\alpha,\beta} F_{U_\alpha \times_S U_\beta !} \rightrightarrows \coprod_{\alpha} F_{U_\alpha !} \to F$$

is acyclic (look at the fibers). Taking an immersion $i: X \to X'$ into a projective prescheme over S, we can take a resolution of the complex $i_! C^*(U, F)$ by object which are f'_*-acyclic ($f': X' \to S$). The spectral sequence of the double complex obtained by applying the functor f'_* yields the expected result.

§ 3. The trace morphism

In this paragraph, we are mainly interested in the morphisms $f: X \to S$ of preschemes which possess the following property: $(S) f$ is a smooth and quasi-projective morphism. The prescheme S is locally noetherian. The dimension d of the fiber at any point $x \in X$ is independent of the considered point.

The number d will be called the relative dimension of X over S.

The sheaf μ_n (n prime to the residual characteristics of S) is defined by the exact sequence:

$$0 \to \mu_n \to \mathbf{G}_m \xrightarrow{n} \mathbf{G}_m \to 0 \qquad (3.0.1)$$

($\mathbf{G}_m$ denotes the sheaf: multiplicative group)

The sheaf on $X: \mu_n^d$ will play the role of a relative orientation sheaf of X over S and will be denoted by $T_{X/S}$. The sheaf $T_{X/S}$ is stable by the change of the base.

Let $S = \operatorname{spec}(k)$, k an algebraically closed field, and X be a complete connected non-singular curve over S. The exact sequence (3.0.1) yields the exact sequence of abelian groups:

$$0 \to H^0(X, \mu_n) \to H^0(X, \mathbf{G}_m) \xrightarrow{n} H^0(X, \mathbf{G}_m) \to H^1(X, \mu_n) \to$$
$$H^1(X, \mathbf{G}_m) \xrightarrow{n} H^1(X, \mathbf{G}_m) \to H^2(X, \mu_n) \to 0. \qquad (3.0.2)$$

Since the field k is algebraically closed, X is complete, and n prime to the characteristic of k, the morphism $H^0(X, \mathbf{G}_m) \xrightarrow{n} H^0(X, \mathbf{G}_m)$ is surjective. Since furthermore the group $H^1(X, \mathbf{G}_m)$ is isomorphic to the Picard's group of X, the sequence (3.0.2) yields two cononical isomorphisms

$$H^1(X, \mu_n) \xrightarrow{\sim} J_n(X)$$

the points of order n of the jacobian variety of X, and

$$\imath_X : H^2(X, \mu_n) \xrightarrow{\sim} \mathbf{Z}/n \tag{3.0.3}$$

Let us suppose now that $X = A_k^1 = \operatorname{spec}(k[t])$, ($k$ algebraically closed field) and let $f: X \to S = \operatorname{spec}(k)$ the canonical morphism. The canonical immersion $i: A_k^1 \to P_k^1$ (projective space of dimension 1 over k) yields the exact sequence of sheaves on P_k^1:

$$0 \to \mu_{nX!} \to \mu_n \to \mu_{n\infty} \to 0$$

Since $\mu_{n\infty}$ is obviously an acyclic sheaf over P_k^1, we get a sequence of isomorphisms:

$$R^2 f_!(\mu_n) \xrightarrow{\sim} H^2(P_k^1, \mu_{nX!}) \xrightarrow{\sim} H^2(P_k^1, \mu_n) \xrightarrow{\sim} \mathbf{Z}/n$$

Let us denote by:

$$\omega_k : R^2 f_!(\mu_n) \xrightarrow{\sim} \mathbf{Z}/n \tag{3.0.4}$$

the composed isomorphism. We can now formulate the main proposition of this paragraph. (From now on, except when explicitly mentioned, the sheaves considered will be sheaves of $\mathbf{Z}/n$-modules).

3.1. Proposition. *It is possible in only one manner to attach to anymorphism $f: X \to S$ satisfying (S), and to any sheaf F on S, one morphism (called the trace morphism):*

$$\varrho_{X,S}(F): R^{2d} f_!(f^*(F) \otimes T_{X/S}) \to F$$

(d is the relative dimension of X over S) such that$_d$

TR0) *$\varrho_{X,S}(F)$ is functorial in F.*

TR1) *$\varrho_{X,S}$ is compatible with the change of the base.*

TR2) *$\varrho_{X,S}$ is compatible with the composition of the morphisms.*

TR3) *When f is étale, $\varrho_{X,S}$ is the canonical morphism yielded by the adjunction formula.*

TR4) *When $S = \operatorname{spec}(k)$, $X = A_k^1$, $F = \mathbf{Z}/n$, the morphism $\varrho_{X,S}(\mathbf{Z}/n)$ is equal to ω_k (3.0.4).*

Furthermore the morphism $\varrho_{X,S}$ possesses the following properties:

(a) *When the fibers of the morphism f are connected and non-empty, $\varrho_{X,S}$ is an isomorphism.*

(b) *When* $S=\mathrm{spec}(k)$ *(k algebraically closed field), when X is a complete, connected, non-singular curve and when F is $\mathbf{Z}/n$, the morphusm $\varrho_{X,S}$ is equal to ι_X* (3.0.3).

(c) *The morphisms $\varrho_{X,S}$ for different n are compatible.*

Let us first elucidate the axiom (TR2). Let $f: X \to S$ and $g: S \to Y$ be two morphisms of preschemes satisfying (S). Let d and d' be the respective relative dimensions. The functor $R^q f_!$ (resp. $R^q g_!$) is null for $q > 2d$ (resp. $q > 2d'$). Therefore the spectral sequence of composition yields an isomorphism

$$R^{2d'} g_! R^{2d} f_! \xrightarrow{\sim} R^{2(d-d')} gf_! .$$

Furthermore the orientation sheaf $T_{X/Y}$ is canonically isomorphic to $T_{X/Y} \otimes f^*(T_{S/Y})$ so that we have, for any sheaf F on Y, a natural isomorphism α that we can include in a diagram:

$$\begin{array}{ccc} R^{2(d+d')} gf_!(T_{X/S} \otimes f^* g^* F) & \xrightarrow[\alpha]{\sim} & R^{2d'} g_! R^{2d} f_!(T_{X/S} \otimes f^*(T_{S/Y} \otimes g^* F)) \\ \Big\downarrow \varrho_{X,Y} & & \Big\downarrow R^{2d'} g_!(\varrho_{X,S}) \\ F & \xleftarrow{\varrho_{Y/S}} & R^{2d'} g_!(T_{S/Y} \otimes g^* F) \end{array} \qquad (3.1.1)$$

The axiom (TR2) is that the above diagram must be commutative.

Proof of the proposition. Uniqueness: By (TR1) we are reduced to the case $S=\mathrm{spec}(k)$ where k is an algebraically closed field. By (TR2), (TR3) and the exact sequence (2.2.1) we are reduced to the case when X is affine and f of the type:

$$X \xrightarrow{g} \mathbf{A}_k^d \to \mathrm{spec}(k)$$

where g is étale and $\mathbf{A}_k^d$ is the affine space of dimension d over k (Definition of smooth morphism). By (TR3) and (TR2) we are reduced to the case $X = \mathbf{A}_k^d$ and $f: \mathbf{A}_k^d \to \mathrm{spec}(k)$ the canonical morphism. By induction on d and (TR2) we are reduced to the case $f: \mathbf{A}_k^1 \to \mathrm{spec}(k)$. Since the functors $R^q f_!$ commute with inductive limits we can suppose that $F = \mathbf{Z}/n$. The axiom (TR4) completes the proof.

Existence. We shall sketch the main steps of the proof.

(1) Suppose that the morphism $\varrho_{X,S}$ is constructed when S and X are affine and when f is of the type $X \xrightarrow{g} \mathbf{A}_S^d \to S$ with g étale and that it satisfies (TRi), $0 \leqslant i \leqslant 4$. Then, by localization on X (2.2.1) and on S, we can construct it in the general case. The properties (TRi), $0 \leqslant i \leqslant 4$, can easily be verified.

(2) There exists one and only one functorial isomorphism:

$$R^{2d}f_!(T_{X/S} \otimes f^* F) \to R^{2d}f_!(T_{X/S}) \otimes F$$

such that he properties (TR1) and (TR2) are satisfied, so that all we have to do is to construct the morphism $\varrho_{X,S}$ only when F is the constant sheaf $\mathbf{Z}/n$.

(3) Suppose that the morphism $\varrho_{X,S}$ is constructed in the two following cases:

(1) The morphism f is étale and the morphism $\varrho_{X,S}$ possesses the properties (TR1), (TR2), and (TR3).

(2) The morphism f is the canonical morphism $\mathbf{A}_S^d \to S$ and the morphism $\varrho_{X,S}$ possesses the properties (TR1), (TR4) and .he property:

(TR2)′ The morphism $\varrho_{X,S}$ is compatible with the S-automorphisms of $\mathbf{A}_S^d$ induced by the permutations of the indeterminates.

Suppose furthermore that the thus constructed morphisms verify the following compatibility property:

(C) For any diagram:

$$\begin{array}{ccc} X & \xrightarrow{g'} & \mathbf{A}_k^1 \\ {\scriptstyle g}\downarrow & & \downarrow{\scriptstyle h'} \\ \mathbf{A}_k^1 & \xrightarrow{h'} & S = \operatorname{spec}(k) \end{array}$$

with g and g' étale and h and h' canonical, the two morphisms $R^2 f_!(T_{X/S} \rightrightarrows \mathbf{Z}/n$ obtained by applying (3.1.1) are equal. Then we can construct $\varrho_{X,S}$ in the general case.

Let us prove this assertion. Let $f: X \xrightarrow{g} \mathbf{A}_S^d \to S$ be a morphism with g étale. We shall define $\varrho_{X,S}$ by the diagram (3.1.1). The only point to be shown is that the so constructed morphism does not depend on the factorization of f. The properties (TRi) can be easily deduced afterward. To show this independence, we can suppose that $S = \operatorname{spec}(k)$ (algebraically closed field). Let us consider

$$\begin{array}{ccc} X & \xrightarrow{g} & A_s^d \\ {\scriptstyle g'}\downarrow & & \downarrow \\ \mathbf{A}_S^d & \to & S = \operatorname{spec}(k) \end{array}$$

two factorizations of f. An S-morphism of X into an affine space over S is determined by d global sections of $0_X: \eta_1, \eta_2, \ldots \eta_d$. Let $\Omega_{X/S}$ be the coherent sheaf of the relative differentials of X on S. The sheaf $\Omega_{X/S}$ is locally free of rank d on 0_X. Let $d\eta_1 \cdots d\eta_d$ be the differentials of the sections $\eta_1 \cdots \eta_d$. The conditions for the morphism g to be étale are that the sections $d\eta_i$ of $\Omega_{X/S}$ generate the sheaf $\Omega_{X/S}$. Let $\eta'_1, \ldots \eta'_d$ be d sections of 0_X which determine the morphism g'. The question being local on X, we can see easily, through permutations of the variables and successive substitutions, that we are reduced to the case $\eta'_i = \eta_i$, $2 \leqslant i \leqslant d$. That means that the following diagram is commutative:

$$\begin{array}{ccc} X & \longrightarrow & \mathbf{A}^1_{A_S^{d-1}} \\ {\scriptstyle g'}\downarrow & & \downarrow \\ \mathbf{A}^1_{A_S^{d-1}} & \to & \mathbf{A}_S^{d-1} \end{array} \qquad S = \operatorname{spec}(k).$$

But now looking at the fibers on $\mathbf{A}_S^{d-1}$ and applying the property (C) we are done.

(4) Let us define the morphism $\varrho_{X,S}$ for f étale in the obvious way. The properties (TR1), (TR2), (TR3) can easily be verified. For $f: \mathbf{A}_S^d \to S$ we shall define $\varrho_{X,S}$ by induction on d so that we are reduced to the case $d=1$. Using arguments similar to those used in the beginning of this paragraph, all that is left for us to do is to define the morphism $\varrho_{X,S}$ when $X = P_S^1$. But in this case the sheaf on S: $R^1 f_! (G_m)$ is canonically isomorphic to the constant sheaf $\mathbf{Z}$ and the construction is easy. The properties (TR1), (TR2)′ and (TR4) are obvious so that we still have to check the property (C). This can be done by classical arguments using the norm.

To achieve the proof, we have to check the properties (a), (b), and (c). The properties (b) and (c) are obvious. To check the property (c) we are immediately reduced to the case $S = \operatorname{spec}(k)$ (algebraically closed field). Then we can use the nice neighborhoods of M. Artin and proceed by induction.

§ 4. Formulation of the duality theorem

In this paragraph the morphism $f: X \to S$ of preschemes with the property (S) will be fixed once for all. The relative dimension of X over S is d.

4.1. The derived category. We shall denote by $D_n(X)$ (resp. $D_n(S)$) the *derived category* of the abelian category of sheaves of $\mathbf{Z}/n$-modules on X (resp. S) [H. S.]. Let us recall briefly what this category is. $D_n(X)$ is the category of complexes $F^\cdot$ of sheaves (the differential is of degree $+1$), up to homotopy in which the morphisms which induce isomorphisms on the objects of cohomology are made isomorphisms [H. S.].

The category $D_n^+(X)$ (resp. $D_n^-(X)$, resp. $D_n^b(X)$) is the subcategory of the complexes $F^\cdot$ of $D_n(X)$ whose objects $(F^\cdot)^l$ are null for $l < l^0(F^\cdot)$, (resp. $l > l^0(F^\cdot)$, resp. $l^0(F^\cdot) < l$ and $l < l^1(F^\cdot)$).

The category $D_n(X)$ possesses a *triangulated structure*, i.e., for any morphism $F^{\cdot} \xrightarrow{u} G^{\cdot}$ we get a triangle that is unique up to non-unique isomorphism (the mapping cylinder):

$$\deg(w) = 1 \qquad \begin{array}{ccc} & H^{\cdot} & \\ {}^{w}\swarrow & & \searrow^{v} \\ F^{\cdot} & \xrightarrow{u} & G^{\cdot} \end{array} \qquad (^*)$$

Such triangles are called the *distinguished triangles.*

A functor $D_n(X) \to D_n(S)$ is *exact* if it transforms distinguished triangles into distinguished triangles.

A *cohomological functor* R from $D_n(X)$ into an abelian category transforms any distinguished triangle (*) into an infinite exact sequence:

$$\cdots \to R^0 F^{\cdot} \to R^0 G^{\cdot} \to R^0 H^{\cdot} \to R^1 F^{\cdot} \to \cdots$$

The usual functor "cohomology" is a cohomological functor with values in the category of sheaves on X.

The functor $\mathrm{Hom}_{(X)}(F^{\cdot}, .)$ (resp. $\mathrm{Hom}_{(X)}(., F^{\cdot})$) is a cohomological functor (resp. a contravariant cohomological functor). The group $\mathrm{Hom}_{(X)}(F^{\cdot}, G^{\cdot})$ is sometimes called the hyper-Ext^0 group.

The category $D_n^+(X)$ is equivalent to the category of complexes of injective sheaves, bounded below, up to homotopy. A *resolution* $F^{\cdot} \to G^{\cdot}$ of a complex $F^{\cdot}$ is a morphism which induces isomorphisms on the cohomology, i.e., which yields an isomorphism in the category $D_n(X)$.

To any sheaf F on X we shall associate the following complex of sheaves on X also denoted by F:

$$\begin{aligned} (F)^l &= 0 \quad \text{for} \quad l \neq 0 \\ (F)^0 &= F \\ d^l &= 0 \end{aligned}$$

The functor thus defined from the sheaves on X into $D_n(X)$ is fully faithful. Exact sequences of sheaves on X yield functorially distinguished triangles.

4.2. The exact functor $\mathbf{R}f_!$. Let $X \xrightarrow{i} X' \xrightarrow{f} S$ be an S-imersion of X into a projective prescheme over S. Let $F^{\cdot}$ be a complex of sheaves on X bounded below and let us take a resolution of $i_! F^{\cdot}$ by a complex of injective sheaves on X'. Applying the functor f'_* we get a complex of sheaves on S and therefore an object of $D_n(S)$ which we shall denote by $\mathbf{R}f_!(F^{\cdot})$. It can be shown that the object $\mathbf{R}f_!(F^{\cdot})$ depends functorially on $F^{\cdot}$, (it does not depend up to unique isomorphism on the injective resolution and on the immersion i, prop. 2.1). Furthermore the functor $\mathbf{R}f_!$ can be uniquely factorized through the category $D_n^+(X)$. The functor thus defined will be again denoted by:

$$\mathbf{R}f_!: D_n^+(X) \to D_n^+(S).$$

The functor $\mathbf{R}f_!$ is exact. For any sheaf P on X the cohomology sheaves of the complex $\mathbf{R}f_!(F)$ are isomorphic to the sheaves $R^q f_!(F)$, (cf. § 2).

Since the functor f'_* is of finite cohomological dimension (2d), the functor $\mathbf{R}f_!$ can be extended to the categories $D_n(X) \to D_n(S)$ (we can take a resolution of any complex on X' by complexes whose objects are f'_*-acyclic) and by restriction to sub-categories yields various functors:

$$\mathbf{R}f_!: D_n^-(X) \to D_n^-(S)$$
$$\mathbf{R}f_!: D_n^b\ X) \to D_n^b(S)$$

4.3. Proposition. 1) *Let $S \xrightarrow{g} Y$ be another morphism of prescheme possessing the property* (S). *The canonical morphism* $\mathbf{R}(gf)_! \to \mathbf{R}g_! \mathbf{R}f_!$ *is an isomorphism.*

2) *Consider the following cartesian square:*

$$\begin{array}{ccc} X & \xleftarrow{u} & X' \\ f\downarrow & & \downarrow f' \\ S & \xleftarrow{u'} & S' \end{array}$$

The canonical morphism of functors:

$$u'^* \mathbf{R}f_! \to \mathbf{R}f'_! u^*$$

is an isomorphism.

The first assertion is obvious, the second one is the base change theorem for proper morphism.

4.4. The twisted inverse image functor. Let $G = \cdots \to G^l \xrightarrow{d^l} G \to G^{l+1} \to \cdots$ be a complex of sheaves on S. We shall denote by $f^!(G^\cdot)$ the following complex of sheaves on X:

$$(f^!(G^\cdot))^l = f^*(G^{l+2d}) \otimes T_{X/S}$$
$$d^l(f^!(G^\cdot)) = f^*(d^{l+2d}) \otimes id_{T_{X/S}}$$

This functor obviously yields an exact functor also denoted by $f^!$:

$$f^!: D_n(S) \to D_n(X)$$

By restriction, this functor yields various functors:

$$D_n^+(S) \to D_n^+(X)$$
$$D_n^-(S) \to D_n^-(X)$$
$$D_n^b(S) \to D_n^b(X)$$

4.5. Proposition 1) *Let $S \xrightarrow{g} Y$ be another morphism of preschemes with the property (S). The canonical morphism $gf^! \to f^! g^!$ is an isomorphism*

2) *Consider the following cartesian square:*

$$\begin{array}{ccc} X & \xleftarrow{u} & X' \\ f\downarrow & & \downarrow f' \\ S & \xleftarrow{u'} & S' \end{array}$$

The canonical morphism $f'^! u'^* \leftarrow u^* f^!$ is an isomorphism:

Those two assertions are obvious.

4.6. The trace morphism in the derived categories. Let $X \xrightarrow{i} X' \xrightarrow{f'} S$ be an S-immersion of X into a projective prescheme over S, and G be a sheaf on S. Let us take now a resolution on X' of the complex $i_! f^!(G)$:

$$0 \to 0 \to I^{-2d} \to I^{-2d+1} \to \cdots \to I^0 \xrightarrow{d^0} I^1 \to \cdots$$

Let Z^0 be the kernel of the morphism d^0. We have a resolution

$$C^*(G) = \cdots 0 \to I^{-2d} \to I^{-2d+1} \to \cdots \to I^{-1} \to Z^0 \to 0 \to \ ,$$

of $i_! f^!(G)$ by f'_*-acyclic objects and therefore the complex on $S: f'_*(C^*(G))$ is canonically isomorphic in $D_n(S)$ to the complex $\mathbf{R}f_! f^!(G)$. But now it is clear that we have a canonical morphism of complexes:

$$\mathbf{R}f_! f^!(G) \to R^{2d} f_!(f^*(G) \otimes T_{X/S})$$

and using the trace morphism we get a functorial morphism:

$$\mathrm{Tr}_{X/S}: \mathbf{R}f_! f^!(G) \to G$$

This morphism can easily be extended (by means of Cartan-Eilenberg resolutions) to any complex of sheaves on S, and yields a morphism of exact functors.

4.7. The duality morphism. Let $K^\cdot$ be an object of $D_n^+(X)$ and $H^\cdot$ be an object of $D_n(X)$. Let us denote by $\mathbf{RHom}(H^\cdot, K^\cdot)$, the following complex on X: Take an injective resolution $I^\cdot$ of the complex $K^\cdot$ and consider the object of $D_n(X)$ defined by the complex of sheaves: $\mathbf{Hom}^*(H^\cdot, I^\cdot)$

where **Hom** is the homomorphism sheaf. Let us assume now that $H^{\cdot}$ is an object of $D_n^-(X)$ and let us apply the functor global section on X. We get now an object of $D(\mathrm{Ab})$ which we shall denote by: $\mathbf{R}\mathrm{Hom}(H^{\cdot}, K^{\cdot})$. The sheaves of cohomology of $\mathbf{RHom}(H^{\cdot}, K^{\cdot})$ are the local hyper-ext. The groups of cohomology of $\mathbf{R}\mathrm{Hom}(H^{\cdot}, K^{\cdot})$ are the global hyper-ext.

In the same way we define $\mathbf{R}f_*(K^{\cdot})$: Take an injective resolution and apply the direct image functor.

By functoriality of $Rf_!$, for any $H^{\cdot}$ object of $D_n^-(X)$ and $K^{\cdot}$ object of $D_n^+(X)$, we get a functorial morphism:

$$\mathbf{R}f_*\mathbf{RHom}(H^{\cdot}, K^{\cdot}) \to \mathbf{RHom}(Rf_!(H^{\cdot}), Rf_!(K^{\cdot}))$$

which gives, when we apply the functor global section on S, a funct orial morphism:

$$\mathbf{R}\mathrm{Hom}(H^{\cdot}, K^{\cdot}) \to \mathbf{R}\mathrm{Hom}(\mathbf{R}f_!(H^{\cdot}), Rf_!(K^{\cdot}))$$

which yields, taking the cohomology, functorial morphism of groups:

$$\mathrm{Ext}^p(H^{\cdot}, K^{\cdot}) \to \mathrm{Ext}^p(\mathbf{R}f_!(H^{\cdot}), \mathbf{R}f_!(K^{\cdot}))$$

But now using the trace morphism, we obtain morphisms:

$$\Delta^1_{X/S}: \mathbf{R}f_*\mathbf{RHom}(F^{\cdot}, f^!G^{\cdot}) \to \mathbf{RHom}(\mathbf{R}f_!F^{\cdot}, G^{\cdot})$$
$$\Delta^2_{X/S}: \mathbf{R}\mathrm{Hom}(F^{\cdot}, f^!G^{\cdot}) \to \mathbf{R}\mathrm{Hom}(\mathbf{R}f_!F^{\cdot}, G^{\cdot})$$
$$\Delta^3_{X/S}: \mathrm{Ext}^p(F^{\cdot}, f^!G^{\cdot}) \to \mathrm{Ext}^p(\mathbf{R}f_!F^{\cdot}, G^{\cdot})$$

for any $F^{\cdot}$ object of $D_n^-(X)$ and $G^{\cdot}$ object of $D_n^+(S)$.

Let us then formulate the duality theorem:

4.8. Theorem (A. Grothendieck). *The morphisms* $\Delta^i_{X/S}$ ($i=1, 2, 3$) are *isomorphisms.*

Remark 1. Assume $S=\mathrm{spec}(k)$ (algebraically closed field) and X connected. Let $G^{\cdot}$ be the group $\mathbf{Z}/n$ and $F^{\cdot}=F$ be a sheaf on X. The duality theorem yields an isomorphism (we shall denote by $H_c^p(X, F)$ the groups $R^pf_!(F)$):

$$\mathrm{Hom}(H_c^p(X, F), \mathbf{Z}/n) \xrightarrow{\sim} \mathrm{Ext}^{2d-p}(X, F, T_{X/S})$$

Assume furthermore that F is locally free and of finite type. Using spectral sequence from local Ext to global Ext we obtain: ($F'=\mathrm{Hom}(F, T_{X/S})$)

$$\mathrm{Hom}(H_c^p(X, F), \mathbf{Z}/n) \xrightarrow{\sim} H^{2d-p}(X, F')$$

which can also be formulated in the following way: The cup-product

$$H_c^p(X, F) \otimes H^{2d-p}(X, F') \to H_c^{2d}X, T_{X/S}) \xrightarrow{\sim} \mathbf{Z}/n$$

is a perfect duality. This is one of the classical formulations of the theorem of Poincaré.

Remark 2. Using localization (associated sheaf) and global section (spectral sequence from local to global), it is easy to see that if for some $i(i=1, 2, 3)$ the morphism $\Delta^i_{X/S}$ is always an isomorphism, all the $\Delta^i_{X/S}$ are isomorphisms.

§ 5. Proof of the duality theorem

We shall sketch the proof of the duality theorem.

Let us recall first that, the preschemes X and S being locally noetherian, the categories of sheaves on X and S are locally noetherian. Let us recall also that a noetherian sheaf G is constructible, i.e., any point possesses a neighborhood which possesses a finite partition into locally closed subsets on which the sheaf G is locally constant and of finite type. In particular any constructible sheaf is locally constant in the neighborhood of the generic point of any irreducible component. It can be shown (as a corollary to the base change theorem for proper morphism) that the direct images (including the q-th direct images $q \neq 0$) of a constructible sheaf by a proper morphism are constructible.

Let $X \xrightarrow{f} S$ be a morphism which possesses the property (S), $F^{\cdot}$ be an object of $D_n^-(X)$ and $G^{\cdot}$ be an object of $D_n^+(S)$. We shall denote by $(i, X, S, F^{\cdot}, G^{\cdot})$ the property: The morphism $\Delta^i_{X/S}(F^{\cdot}, G^{\cdot})$ is an isomorphism.

5.1. Lemma. *The following properties are equivalent:*

(i) *The duality theorem is true for the morphism f.*

(ii) *There exists an* $i\,(i=1, 2, 3)$ *and an étale covering* $(U_j \to X)$ *such that, for any quasi-projective étale morphism $U \to X$ which can be factorized by the above covering and for any constructible sheaf G on S we have the property* $(i, X, S, \mathbf{Z}/n_{U!}, G)$.

5.2. Lemma. *The duality theorem is true when f is étale.*

5.3. Lemma. *Let $S \xrightarrow{f} Y$ be a morphism with the property (S), $H^{\cdot}$ an object of $D_n^+(Y)$ and i be an integer $0 < i \leqslant 3$. Let us suppose that two of the properties below hold:*

$$(i, X, S, F^{\cdot}, g^! H^{\cdot}) \quad (i, X, Y, F^{\cdot}, H^{\cdot}) \quad (i, S, Y, Rf_! F^{\cdot}, H^{\cdot})$$

Then the third one also holds.

Proof. The last lemma comes directly from the transitivity property (4.3; 4.5; 3.1 (TR 2)). The second one is obvious. Let us prove the first one. Any object $F^{\cdot}$ of $D_n^-(X)$ admits a resolution $P^{\cdot}\to F^{\cdot}$ by a complex of the type $:P^{\cdot}=\cdots\to\mathbf{Z}/n_{U_i!}\to\mathbf{Z}/n_{U_k!}\to 0\to\cdots$ where the U_i and the U_k are étale over X and can be factorized by the covering $(U_j\to X)$. So that, by spectral sequence argument or by the way out functor lemma [H. S.] in order to prove the duality theorem, we are brought back to the case $F^{\cdot}=\coprod_i \mathbf{Z}/n_{U!}$, with U_i noetherian. Since the functors $R^q f_!$ and Ext^p commute with the infinite sum we are brought back to the case $F^{\cdot}=\mathbf{Z}/n_{U_i!}$. Again by spectral sequence argument we can suppose that $G^{\cdot}=G$ is one sheaf over S. But now, since the sheaves $\mathbf{Z}/n_{U_i!}$ and $R^q f_!(\mathbf{Z}/n_{U_i!})$ are noetherian sheaves, the hyper-ext $\mathrm{Ext}^*(\mathbf{R}f_!(\mathbf{Z}/n_{U_i!}), G)$ and $\mathrm{Ext}^*(\mathbf{Z}/n_{U_i!}, f^!G)$ commute with the direct limit of G and therefore we can suppose that G is noetherian, i.e., constructible. We thus prove the implication (ii)$\Rightarrow$(i). The other implication is obvious.

5.4. First reduction. *Using the three lemmas above and straight forward arguments we are brought back to the proof of the theorem in the following case:*

(a) *The morphism $f:X\to S$ is of relative dimension 1. X and S are affine noetherian.*

(b) *The complex $F^{\cdot}$ is the constant sheaf $\mathbf{Z}/n$.*

(c) *The complex $G^{\cdot}$ is a constructible sheaf.*

Thus, by the first reduction, we have to check that the morphism

$$\Delta^1_{X/S}: \mathbf{R}f_*(f^!G)\to \mathbf{RHom}(\mathbf{R}f_!(\mathbf{Z}/n), G) \tag{5.4.1}$$

is an isomorphism.

Let $\eta\in S$ be a generic point of an irreducible component of S. The sheaves G and $R^q f_!(\mathbf{Z}/n)$ are constant on an étale neighborhood of η (they are constructible). Therefore the cohomology sheaves of the complex $\mathbf{RHom}(\mathbf{R}f_!(\mathbf{Z}/n), G)$ are constant on an étale neighborhood of η.

5.5. Lemma. *Assume conditions* (a), (b), (c) *of the reduction* 5.4. *Denote by $f_{\bar\eta}$ a geometric fiber of f on a geometric point $\bar\eta$ over η. The cohomology sheaves of $\mathbf{R}f_*(f^!G)$ are constant on an étale neighborhood of η. The complex $\mathbf{R}f_*(f_!G)_{\bar\eta}$ is canonically isomorphic (in $D_n(\mathrm{Ab})$) to the complex $\mathbf{R}f_{\bar\eta,*}(f^!_{\bar\eta}G_{\bar\eta})$.*

We shall not prove this lemma. It follows from the "relative purity theorem" [S. G. A. A.], which is one of the consequences of 1.1.

But now, by the lemma 5.5, to see that the morphism (5.4.1) is an isomorphism on a neighborhood of η, it is enough to look at the fiber $X\bar\eta\to\bar\eta$, i.e., we are reduced to the case $S=\mathrm{spec}(k)$ (k an algebraically closed field). Furthermore, to prove that (5.4.1) is an isomorphism we are reduced, by an easy noetherian induction on the support of the sheaf G, to the case where the support of G is a closed point of S, and we are immediately reduced again to the case $S = \operatorname{spec}(k)$ (algebraically closed field).

Let us suppose that $S = \operatorname{spec}(k)$, we can embed the curve X into a complete non-singular curve X' and we are easily reduced to prove the duality theorem in the case

(a)′ $X \to S$ is a complete non-singular curve over an algebraically closed field.

(b)′ The complex $F^{\cdot}$ is the constant sheaf $\mathbf{Z}/n$.

(c)′ The complex $G^{\cdot}$ is the constant sheaf $\mathbf{Z}/n$.

Thus we have to prove that the morphisms:

$$H^0(X, \mu_n) \to \operatorname{Hom}(H^2(X, \mathbf{Z}/n), \mathbf{Z}/n) \tag{5.5.1}$$

$$H^1(X, \mu_n) \to \operatorname{Hom}(H^1(X, \mathbf{Z}/n), \mathbf{Z}/n) \tag{5.5.2}$$

$$H^2(X, \mu_n) \to \operatorname{Hom}(H^0(X, \mathbf{Z}/n), \mathbf{Z}/n) \tag{5.5.3}$$

are isomorphisms. This can be seen easily for (5.5.1) and (5.5.3). For (5.5.2) this follows from the autoduality of the jacobian variety of X.

References

ARTIN, M. and A. GROTHENDIECK: Séminaire de géométrie algébrique de l'Institut des Hautes Etudes Scientifiques (1964).

ARTIN, H.: Grothendieck Topologies, Notes on a seminar. Harvard University 1962.

HARTSHORNE, R.: Residues and Duality. Lecture Notes in Mathematics 20. Berlin – Heidelberg – New York: Springer 1966.

The Lefschetz Fixed Point Formula in Etale Cohomology

J. L. VERDIER

1. The Lefschetz theorem for a curve

Our aim is to give some indication concerning the proof of

Theorem 1.1. *Let X be a complete non singular curve over an algebraically closed field k. Let L be an étale sheaf of Q_l-vector spaces (l prime to* char(k)*). We assume that L is constructible. This implies that the groups $H^q(X, L)$ are finite dimensional vector spaces. Let*

$$f: X \to X$$

be a morphism and

$$\Phi: f^*(L) \to L$$

a lifting of the morphism f to the sheaf L. The pair (f, Φ) yields maps

$$(f, \varphi)^q : H^q(X, L) \to H^q(X, L)$$

Assume now that f has only isolated fixed points (i.e., $f \neq id_X$). *At each fixed point P we can consider the map induced by Φ on the stalk L_P*

$$\varphi_P : L_P \to L_P .$$

Since the stalks are finite dimensional vector spaces the morphism φ_P possesses a trace.

Assume now that all the fixed points are of multiplicity one. Then we have

$$\sum_q (-1)^q \operatorname{Tr}(f, \varphi)^q = \sum_P \operatorname{Tr}(\varphi_P).$$

Remark. 1) It is not clear what we mean by constructible sheaf of Q_l-vector space and by $H^q(X, L)$. This point will be discussed later on. But let me say only that the definitions of those objects are not the naïve ones.

2) A. GROTHENDIECK has now a shorter proof of this theorem. He even

has a formula which includes the case of arbitrary fixed points and also a formula for the case $f=$ identity. However, the method of Grothendieck applies only to the case of curves. We will present a proof which uses a general Lefschetz formula which, roughly speaking, says that the left hand side is equal to a sum of purely local terms. In the case of a curve we will be able to compute those local terms.

2. The duality theorems

To prove the theorem we need to recall some facts on duality. There are three types of duality theorems we will use. The first two types are formal and quite general. The third one is not at all formal (Poincaré duality) and uses strongly the geometrical assumptions.

I. Let $X \xrightarrow{f} Y$ be a morphism of schemes, F a sheaf on Y (for the étale topology), G a sheaf on X. Then we have an isomorphism

$$R\,\mathrm{Hom}(F, Rf_*G) \xrightarrow{\sim} R\,\mathrm{Hom}(f^*F, G). \tag{2.1}$$

This formula means simply this: Take an injective resolution of G, and apply the functor direct image. You obtain thus a complex Rf_*G. Take now the complex of morphisms of F into Rf_*G. You obtain thus $R\,\mathrm{Hom}$ (F, Rf_*G). On the other hand, take the complex of morphisms of f^*F (inverse image of F) into an injective resolution of G. You obtain the complex $R\,\mathrm{Hom}(f^*F, G)$. The formula says that there exists a natural morphism between those two complexes (well defined up to homotopy) which induces an isomorphism on the cohomology groups.

We can also introduce a ring A and work with sheaves of A-modules. We will take then injective resolutions and morphisms in the category of sheaves of A-modules. This formula has nothing to do with schemes and étale topology. It is a general formula on categories of sheaves on topologies. It is a cohomological translation of the adjointness properties of the functors f^* and f_* (see [1] and [3]). We can also generalize the formula to complexes F and G (see [2]).

II. Let X be a scheme and F, G, H three abelian sheaves (for the étale topology). We have

$$R\,\mathrm{Hom}(F \overset{L}{\otimes} G, H) \xrightarrow{\sim} R\,\mathrm{Hom}(F, R\,\mathscr{H}\mathrm{om}(G, H)). \tag{2.2}$$

The formula means this: Take an injective resolution $I(H)$ of H and a resolution $P(G)$ of G of the form

$$\cdots \to \bigoplus_j \mathbb{Z}_{U_j} \to \bigoplus_i \mathbb{Z}_{U_i} \to G,$$

(what is really important is to take a resolution of G by flat sheaves), where $U_i \to X$ are étale schemes over X and where $\mathbb{Z}_{U_i}$ is the constant sheaf $\mathbb{Z}$ on U_i extended by zero elsewhere. We can now consider $F \overset{L}{\otimes} G = F \otimes P(G)$ (tensor product of complexes) and the complex of morphisms of $F \overset{L}{\otimes} G$ into the complex $I(H)$ which is the complex that we denote by

$$R\,\mathrm{Hom}(F \overset{L}{\otimes} G, H).$$

We can also consider the complex of sheaves of morphisms

$$\mathscr{H}om(P(G), I(H)),$$

which is an injective complex that we denote by $R\mathscr{H}om(G, H)$ and again the complex $R\,\mathrm{Hom}(F, R\mathscr{H}om(G, H))$ of morphisms of F into the complex $R\mathscr{H}om(G, H)$. As before, the formula says that there exists a natural map between the two complexes which induces an isomorphism on the cohomology groups.

As for type I, this formula is also quite general and can be generalized to the case of sheaves of A-modules and also to complexes of such sheaves.

III. Let $X \xrightarrow{f} Y$ be a smooth morphism and assume that f can be factorized by

$$\begin{array}{ccc} X & \xrightarrow{i} & X' \\ & {\scriptstyle f}\searrow \quad \swarrow{\scriptstyle f'} & \\ & Y & \end{array}$$

where i is an immersion and f' is proper. We will assume also, to simplify, that the dimension of X over Y, which is locally constant on X, is in fact constant $=d$. Let n be an integer prime to the residual characteristics of Y. Let $\mathscr{T}_{X/Y}$ be the *orientation complex*, i.e.:

$$\begin{array}{lccccc} \text{object:} & 0 & \to \mu_n^{\otimes d} \to & 0 & \to \cdots \to 0 \to \cdots. \\ \text{degree:} & -2d-1 & -2d & -2d+1 & \cdots, \end{array}$$

where μ_n is the sheaf of nth roots of unity, i.e., the kernel of the morphism

$$G_m \xrightarrow{n} G_m \to 0.$$

In those conditions we can define:

a) For any sheaf F on X (more generally for any complex on X) of $\mathbb{Z}/n$

modules, the complex:

$$R_! f(F)$$

(direct image with proper supports) in the following way:

Let $F_!$ be the sheaf on X′ obtained by extending F by zero. Take an injective resolution of $F_!$ on X'. Apply the functor f'_*. It can be shown that the complex $R_! f(F)$ thus obtained does not depend, up to unique isomorphism in the derived category of complexes of sheaves on Y, on the factorization.

b) For any sheaves G on Y (more generally for any complex of sheaves) of $\mathbb{Z}/n$-modules, the complex

$$f^!(G) = f^*(G) \otimes \mathcal{T}_{X/Y}.$$

Theorem 2.3. *We have, when F and G are n-torsion sheaves:*

$$R\,\mathrm{Hom}(R_! f(F), G) \xrightarrow{\sim} R\,\mathrm{Hom}(F, f^!(G)).$$

(When F and G are complexes we must assume that the cohomology of F is bounded above and that the cohomology of G is bounded below.) As before, $R\,\mathrm{Hom}(\,)$ meant the complex of morphisms into an injective resolution in the category of n-torsion sheaves.

There is also a Poincaré duality theorem for quasi-finite morphisms. Let $f: X \to Y$ be a quasi-finite morphism and

$$\begin{array}{ccc} X & \subsetneq & X' \\ {\scriptstyle f}\searrow & & \swarrow{\scriptstyle f'} \\ & Y & \end{array}$$

a factorization through a finite morphism.
For any sheaf F on X let $F_!$ be the extension by zero on X' and

$$f_! F = f'_* F_! .$$

When f is étale, then $f_!$ is simply the extension by zero. When f is a closed immersion, then $f_!$ is simply the direct image.

Let G be a sheaf of n-torsion on Y (n invertible on Y) and $I(G)$ an injective resolution of G. For any $U \to X$ étale set:

$$f^!(G)(U) = \mathrm{Hom}(f_1(\mathbb{Z}/n_U), I(G)).$$

$U \rightsquigarrow f^!(G)(U)$ is a sheaf on X that we denote by $f^!(G)$. When f is étale, $f^!$ is simply the restriction functor. When f is a closed immersion, $\mathcal{H}^0 f^!(G)$ is the functor "sections of G whose supports are in X", and the $\mathcal{H}^i f^!(G)$

are the derived functors of the functor $\mathscr{H}^0 f^!(G)$. The direct images on Y will be denoted by

$$\mathscr{H}^i_X(Y, G).$$

Theorem 2.4. $R\,\mathrm{Hom}(f_!(F), G) \xrightarrow{\sim} R\,\mathrm{Hom}(F, f^!(G))$.

The proof is immediate. This theorem, when f is a closed immersion, can be extended to a general topos. (cf. [*3*]).

3. Four fundamental properties

Let X be a smooth complete scheme over an algebraically closed field k. Let F be a sheaf on X of n-torsion (n prime to $\mathrm{char}(k)$) and $f: X \to X$, $\varphi: f^* F \to F$ a morphism of (X, F) into (X, F). To prove a Lefschetz theorem, we need to check 4 fundamental properties.

(1) Finiteness conditions on F.

(2) Poincaré duality theorem for the morphisms $X \to \mathrm{spec}(k)$, $X \times X \xrightarrow{pr_1} X$, $X \xrightarrow{\Gamma} X \times X$. ($\Gamma$ is the graph morphism.)

(3) Künneth Formula.

(4) Biduality property for the sheaf F.

The finiteness conditions will be fulfilled when F is a constructible sheaf. We have already recalled that we have a Poincaré duality theorem. We will show in this paragraph that the Künneth formula holds. We will also explain the biduality property and state that a large class of sheaves on X possesses, in some cases, the biduality property.

Theorem 3.1. *Consider the following situation: f and g*

$$\begin{array}{ccccc} & & X \underset{Z}{\times} Y & & \\ & {\scriptstyle p}\swarrow & & \searrow{\scriptstyle q} & \\ F, X & & \downarrow{\scriptstyle h} & & Y, G \\ & {\scriptstyle f}\searrow & & \swarrow{\scriptstyle g} & \\ & & Z & & \end{array}$$

are "compactifiable" morphisms (i.e., *which factorize through an immersion and a proper morphism*), *F and G are n-torsion sheaves (n invertible on Z). Then we have a canonical isomorphism:*

$$R_! f(F) \overset{L}{\otimes} R_! g(G) \xrightarrow{\sim} R_! h(p^*(F) \overset{L}{\otimes} q^*(G)).$$

Corollary 3.2. *Let $g = \text{identity}$. We have:*

$$R_! f(F) \underset{L}{\overset{L}{\otimes}} G \underset{m}{\xrightarrow{\sim}} R_! f(F \underset{L}{\overset{L}{\otimes}} f_!^*(G)).$$

In those formulas, the $\overset{L}{\otimes}$ means tensor product with a flat resolution in the category of n-torsion sheaves.

Proof. The theorem follows from the Corollary. (One uses the base change theorem.) To prove the corollary one can assume that f is prover. First we have to define a canonical morphism. Take a finite f_* acyclic resolution $R(F)$ of F, a flat resolution $P(G)$ of G and a resolution of the complex $R(F)\otimes P(G)$ by a complex of f_* acyclic objects bounded above. (This is possible, because the functor f_* is of finite cohomological dimension.) Applying the functor f_* we obtain a morphism

$$Rf_*(F)\overset{L}{\otimes} G \overset{m}{\to} Rf_*(F\overset{L}{\otimes} f^*(G)).$$

By the spectral sequence, we are reduced (to prove that m is an isomorphism) to the case $G=\bigoplus_i \mathbb{Z}/n_{U_i}$, where $U_i\to Z$ are étale. Using the fact that the $R^q f_*$ commute with direct limits, we are reduced to the case $G=\mathbb{Z}/n_U$. But now if suffices to look at the fibers (f is proper) to complete the proof.

Let X be a smooth variety over a field k algebraically closed:

$$\begin{array}{c} X \\ f\downarrow \\ \operatorname{spec}(k) \end{array}$$

Let $\mathcal{T}_X$ be the orientation complex (for n-torsion sheaves), let F be a finite complex whose cohomology is n-torsion on X. The complex

$$D_X F = R\mathcal{H}om(F, \mathcal{T}_X)$$

will be called the dual complex of F. It is easily seen that $D_X F$ is again a finite complex whose cohomology is n-torsion. Instead of $H^i R_! f$, we will use the classical notation $H_c^i(X, \cdot)$. The duality theorem for F says (n prime to char(k)):

$$\operatorname{Hom}(H_c^q(X, F), \mathbb{Z}/n) \xrightarrow{\sim} H^{-q}(X, D_X F),$$

i.e., the dual of the cohomology with compact support is canonically isomorphic to the cohomology of the dual.

We will say that the complex F possesses the (weak) biduality property if the canonical morphism

$$F \to D_X D_X F$$

is an isomorphism (in the derived category).

Obviously if F is locally free of finite rank, F possesses the weak biduality property. It is also the case when F is locally constant of finite type. It is clear that we have to make some finiteness assumptions on F to have a biduality property.

Definition 3.3. *A sheaf F on a locally noetherian scheme S is said to be constructible if each point $x \in S$ possesses a neighborhood V_x and a partition*

$$V_x = \bigcup_i Z_i$$

into locally closed subsets such that on each Z_i (say with the reduced structure) the sheaf F induces a locally constant sheaf of finite type.

It can be shown that the constructible sheaves on S (noetherian) are exactly the noetherian objects in the category of n-torsion sheaves ([3]).

Let Y be another smooth variety over k and G be an n-torsion sheaf over Y. It is clear that we have a canonical isomorphism

$$\begin{array}{ccc} & X \times Y & \\ {}^{pr_1}\swarrow & & \searrow{}^{pr_2} \\ {}_F X & & Y_G \end{array}$$

$$\mathcal{T}_{X \times Y} \xrightarrow{\sim} \mathcal{T}_X \otimes \mathcal{T}_Y$$

and therefore, using duality formulas of type II, we have a canonical morphism:

$$pr_1^* D_X F \overset{L}{\otimes} pr_2^* G \to D_{X \times Y}(pr_1^* F \overset{L}{\otimes} pr_2^* D_Y G). \qquad (*)$$

The quadrupole (X, Y, F, G) is said to possess the biduality property if (*) is an isomorphism.

A sheaf F on X is said to possess the strong biduality property if F possesses the weak biduality property and if the quadruple (X, X, F, F) possesses the biduality property.

A locally free sheaf of finite type possesses the strong biduality property.

We will say that *one can resolve the singularities* on X if, given any étale morphism: $U \xrightarrow{\pi} X$ of finite type, there exists a factorization

$$\begin{array}{ccc} U & \overset{i}{\hookrightarrow} & X' \\ {}_{\pi}\downarrow & \swarrow_h & \\ X & & \end{array}$$

where i is an open immersion, h proper, X' smooth over k, and $X' - U$ is a union of divisors which cross normally.

Theorem 3.4. *Assume that one can resolve the singularities on X and Y. For any constructible sheaf F on X whose fibers are free $\mathbb{Z}/n$-modules and*

G on Y, the quadruple (X, Y, F, G) possesses the biduality property. Moreover the source of (*) is a constructible sheaf.

Corollary 3.5. *Assume that one can resolve the singularities on X. Any constructible sheaf F on X whose fibers are free $\mathbb{Z}/n$-modules possesses the strong biduality property. Moreover DF is also constructible.*

The next proposition will be the key result in the proof of the Lefschetz theorem.

Proposition 3.6. *Let X be a smooth variety over k, F a sheaf which possesses the strong biduality property. Then,*

$$\begin{array}{ccc} & X \times X & \\ {\scriptstyle pr_1}\swarrow & & \searrow{\scriptstyle pr_2} \\ X & & X \end{array}$$

we have on $X \times X$, a canonical isomorphism (in the derived category):

$$R\mathscr{H}\text{om}(pr_1^*F, pr_2^!F) \xrightarrow{\sim} pr_1^*D_XF \overset{L}{\otimes} pr_2^*F.$$

Proof. We have a biduality isomorphism:

$$F \xrightarrow{\sim} D_XD_XF$$

and therefore

$$R\mathscr{H}\text{om}(pr_1^*F, pr_2^!D_XD_XF) \xleftarrow{\sim} R\mathscr{H}\text{om}(pr_1^*F, pr_2^!F).$$

But, as an easy consequence of the duality theorem for the morphism pr_2, we have an isomorphism (for any sheaf G on X):

$$pr_2^!D_XG \xrightarrow{\sim} D_{X\times X}pr_2^*G.$$

Hence

$$R\mathscr{H}\text{om}(pr_1^*F, pr_2^!F) \xrightarrow{\sim} R\mathscr{H}\text{om}(pr_1^*F, D_{X\times X}pr_2^*D_XF)$$

Using a formula of type II we obtain

$$R\mathscr{H}\text{om}(pr_1^*F, pr_2^!F) \xrightarrow{\sim} D_{X\times X}(pr_1^*F \overset{L}{\otimes} pr_2^*D_XF),$$

and using the strong biduality property we complete the proof.

4. Sheaves of Q_l-vector spaces

Let X be a locally noetherian scheme and l a prime. A $\mathbb{Z}_l$-sheaf on X is an inverse system

$$\cdots F_n \to F_{n-1} \to \cdots$$

such that F_n is a l^n-torsion sheaf and such that the transition morphism

$$F_n \to F_{n-1}$$

is isomorphic to the morphism

$$F_n \to F_n \underset{\mathbb{Z}/l^n\mathbb{Z}}{\otimes} \mathbb{Z}/l^{n-1}\mathbb{Z}$$

The $\mathbb{Z}_l$-shaves on X form a category and even better a $\mathbb{Z}_l$-category, i.e., there exists a canonical map of the ring $\mathbb{Z}_l$ into the ring of endomorphisms of the identity functor.

A $\mathbb{Z}_l$-sheaf is constructible iff all the F_n are constructible. In the category of constructible $\mathbb{Z}_l$-sheaves, we can define a subcategory of torsion objects. The quotient category is the category of constructible sheaves of Q_l vector spaces.

The basic example of constructible sheaf of Q_l-vector spaces is given by a continuous representation

$$\pi_1(X, a) \to GL(V)$$

of the π_1 into the group of automorphisms of a finite dimensional Q_l-vector space.

We will admit here that the theorem 1.1 can be proved by going up to the limit on n-torsion constructible sheaves, and we will restate the theorem 1.1 for torsion sheaves. It is however necessary to define the trace, which we will do now.

Let A be a noetherian ring. Let us denote by $D_f^b(A)$ the full subcategory of the derived category of the category of A-modules defined by the complexes isomorphic to finite complexes of projective modules of finite type. A complex $L \in D(A)$ is an object of $D_f^b(A)$ iff:

(a) $H^q(L)$ is a module of finite type for any q and $H^q(L)=0$ for $|q|$ big enough.

(b) There exists a q_0 such that for any module M

$$H^q(L \overset{L}{\otimes} M) = 0, \qquad \forall q < q_0 .$$

Let $\psi: L \to L$ be an endomorphism of an object of $D_f^b(A)$. The morphism ψ is isomorphic to an endomorphism

$$\psi': L' \to L'$$

where L' is a finite complex of projective modules of finite type. Let

$\psi'_q: L'_q \to L'_q$ be the "components"[1] of ψ' and set

$$\mathrm{Tr}(\psi) = \sum_q (-1)^q \,\mathrm{Tr}\,\psi_q$$

where in the right hand side the traces are the usual ones. One checks easily that this definition makes sense and that this trace function possesses the following properties:

(TR1): For any distinguished triangle of endomorphisms:

$$\begin{array}{ccc} & \psi_3 & \\ {\scriptstyle \deg 1}\swarrow & & \nwarrow \\ \psi_1 & \to & \psi_2 \end{array}$$

We have;

$$\mathrm{Tr}\,\psi_1 - \mathrm{Tr}\,\psi_2 + \mathrm{Tr}\,\psi_3 = 0.$$

(TR2): For any diagram: $X \xrightarrow{m} Y \xrightarrow{n} X$ we have

$$\mathrm{Tr}(m_0 n) = \mathrm{Tr}(n_0 m).$$

(TR3): $$\mathrm{Tr}(\psi_1 \overset{L}{\otimes} \psi_2) = \mathrm{Tr}(\psi_1)\,\mathrm{Tr}(\psi_2).$$

Let now X be a complete algebraic scheme over an algebraically closed field k, and let F be a constructible n-torsion sheaf on X (n prime to char(k)). Denote by $RH(X, F)$ the complex of global sections on X. This complex $RH(X, F)$ possesses the property (a) above ([3]). Assume now that each fiber of F is flat (say, in this case that each fiber of F is a free $\mathbb{Z}/n$ module of finite rank). The Corollary 3.2 shows that the complex $RH(X, F)$ possesses also the property (b) above. Therefore any endomorphism

$$\psi: RH(X, F) \to RH(X, F)$$

possesses a trace.

We can now restate the theorem 1.1 for n-torsion sheaves.

Theorem 4.1. *Let X be a complete non singular curve over an algebraically closed field k. Let F be a constructible n-torsion sheaf whose fibers are free $\mathbb{Z}/n$-modules. Let*

$$f: X \to X \quad \varphi: f^* F \to F$$

be an endomorphism of (X, F).

Assume that f possesses only isolated fixed points and that each fixed point is of multiplicity one. We can then define $\chi(f, \varphi, F)$, the trace of the

[1] They are defined only "up to homotopy".

morphism

$$\psi: RH(X, F) \to RH(X, F)$$

induced by (f, φ) *and at each fixed point* $P \in X$, $\mathrm{Tr}_p(\Phi)$ *the trace of the morphism* Φ *induced on the fiber. We have*

$$\chi(f, \varphi, F) = \sum_P \mathrm{Tr}_p(\Phi).$$

5. Lefschetz theorem

Let us consider the Lefschetz situation: X is complete and smooth over k algebraically closed, F is an n-torsion constructible sheaf whose fibers are free (n prime to $\mathrm{char}(k)$) and which possesses the strong biduality property, $f: X \to X$, $\varphi: f^*(F) \to F$, an endomorphism of (X, F). We therefore have a diagram:

$$\begin{array}{ccc} & X & \\ & \downarrow \Gamma & \\ & X \times X & \\ pr_1 \swarrow & & \searrow pr_2 \\ X & & X \end{array} \qquad \begin{array}{l} pr_1 \circ \Gamma = f \\ pr_2 \circ \Gamma = id_X \end{array}$$

and a morphism:

$$\varphi: \Gamma^* pr_1^* F \to F.$$

The equality $pr_2 \Gamma = id_X$ implies $\Gamma^! pr_2^! \simeq$ identity and thus φ defines a morphism

$$\varphi: \Gamma^* pr_1^* F \to \Gamma^! pr_2^! F.$$

One checks easily that φ defines an element

$$\alpha \in H^0_\Gamma(X \times X, R\mathscr{H}\mathrm{om}(pr_1^* F, pr_2^! F))$$

and therefore an element

$$\beta \in H^0(X \times X, R\mathscr{H}\mathrm{om}(pr_1^* F, pr_2^! F)).$$

Now using the proposition 3.6 we obtain an element

$$\eta \in H^0(X \times X, pr_1^* DF \overset{L}{\otimes} pr_2^* F)$$

and we can consider the restriction of η to the diagonal

$$\eta_\Delta \in H^0(X, DF \overset{L}{\otimes} F).$$

But there is a canonical pairing $DF \overset{L}{\otimes} F \to \mathscr{T}_X$ and a canonical morphism

(the fundamental class)

$$H^0(X, \mathcal{T}_X) \to \mathbb{Z}/n$$

which yields the residue map:

$$\varrho: H^0(X, DF \overset{L}{\otimes} F) \to \mathbb{Z}/n .$$

Theorem 5.1. $\chi(f, \Phi, F) = \varrho(\eta_\Delta)$.

Proof. Let $\psi: RH(X, F) \to RH(X, F)$ be the morphism defined by (φ, f). The complex $RH(X, F)$ is isomorphic to a finite complex of free $\mathbb{Z}/n$-modules of finite rank. Therefore, ψ defines an element

$$\alpha' \in H^0(RH(X, F)^* \overset{L}{\otimes} RH(X, F))$$

where the * means the complex of homomorphisms into $\mathbb{Z}/n$. The trace of ψ is $\varrho'(\alpha)$ where

$$\varrho': H^0(RH(X, F)^* \overset{L}{\otimes} RH(X, F)) \to \mathbb{Z}/n$$

is the canonical pairing.

The Poincaré duality theorem on X says that we have a canonical isomorphism

$$RH(X, F)^* \xrightarrow{\sim} RH(X, DF),$$

and that the canonical pairing

$$H^0(RH(X, F)^* \overset{L}{\otimes} RH(X, F)) \to \mathbb{Z}/n$$

is, modulo this isomorphism, defined by the cup product

$$RH(X, DF) \overset{L}{\otimes} RH(X, F) \to RH(X, DF \overset{L}{\otimes} F)$$

and the residue morphism

$$\varrho: H^0(RH(X, DF \overset{L}{\otimes} F)) \to \mathbb{Z}/n .$$

But, as usual, the cup product can be computed via the cross product

$$RH(X, DF) \overset{L}{\otimes} RH(X, F) \to RH(X \times X, pr_1^* DF \overset{L}{\otimes} pr_2^* F)$$

and the restriction to the diagonal

$$RH(X \times X, pr_1^* DF \overset{L}{\otimes} pr_2^* F) \to RH(X, DF \overset{L}{\otimes} F).$$

Let

$$\eta' \in H^0(X \in X, pr_1^* DF \overset{L}{\otimes} pr_2^* F)$$

be the element defined by α'. We have just shown that

$$\chi(f, \varphi, F) = \varrho(\eta'_\Delta)$$

and the theorem follows from

Lemma 5.2.
$$\eta = \eta' .$$

To prove this lemma we just have to check that all the canonical isomorphisms introduced are compatible. We will not go into the details.

Assume now that f has only isolated fixed points. The cohomology class $\eta \in H^0(X \times X, pr_1^* DF \overset{L}{\otimes} pr_2^* F)$ comes, in fact, from a cohomology class

$$\xi \in H^0_\Gamma(X \times X, pr_1^* DF \overset{L}{\otimes} pr_2^* F).$$

Hence the class

$$\eta_\Delta \in H^0(X, DF \overset{L}{\otimes} F)$$

comes, in fact, from a cohomology class

$$\xi_\Delta \in H^0_{\bigcup_i P_i}(X, DF \overset{L}{\otimes} F) \simeq \bigoplus_i H^0_{P_i}(X, DF \overset{L}{\otimes} F),$$

where P_i are the fixed points.

It follows easily, from the Poincaré duality theorem that we have an isomorphism:

$$H^0_{P_i}(X, \mathscr{T}_X) \xrightarrow{\sim} \mathbb{Z}/n,$$

such that the diagram

$$\begin{array}{ccc} \bigoplus_i H^0_{P_i}(X, \mathscr{T}_X) & \to & H^0(X, \mathscr{T}_X) \\ \downarrow & & \downarrow \\ \bigoplus_i \mathbb{Z}/n & \xrightarrow{\Sigma} & \mathbb{Z}/n \end{array}$$

is commutative.

Let us denote by ϱ_{P_i} the morphism

$$\bigoplus H^0_{P_i}(X, DF \overset{L}{\otimes} F) \to H^0_{P_i}(X, DF \overset{L}{\otimes} Ff \to H^0_{P_i}(X, \mathscr{C}_X) \xrightarrow{\sim} \mathbb{Z}/n .$$

We have

Theorem 5.3.
$$\chi(f, \varphi, F) = \sum_i \varrho_{P_i}(\xi_\Delta) .$$

But now the ϱ_i are purely local terms, i.e., to compute $\varrho_i(\xi_\Delta)$ we only have to know the class ξ_Δ in an étale neighborhood of the fixed point P_i.

6. Proof of the Theorem 4.1

It remains to prove that, when the fixed point P is of multiplicity one, we have:

$$\varrho_P(\xi_\Delta) = \mathrm{Tr}(\varphi_P).$$

By an easy devissage, we are reduced to prove

Proposition 6.1. *If P is of multiplicity one, and if the fiber F_P is zero, then*

$$\varrho_P(\xi_\Delta) = 0.$$

This proposition has been proved by M. ARTIN. We will only give some indications in a simple case.

First, the question is purely local. Let $X \times X_p$ (resp. X_p) denote the scheme strictly localized at the point (P, P) (resp. P). We have the following picture:

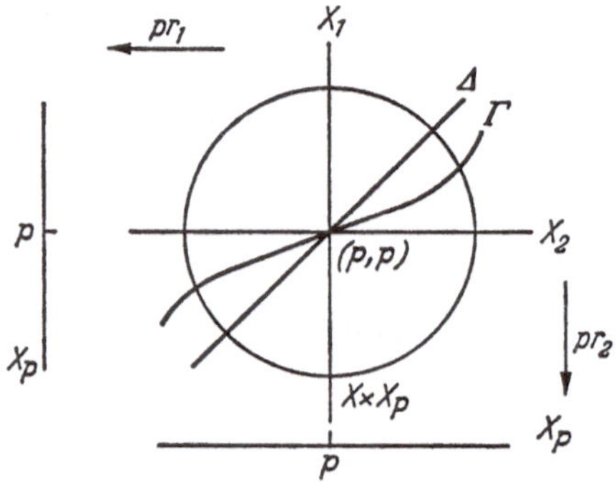

To prove the proposition, it suffices to prove that the restriction map

$$r': H^0_\Gamma(X \times X_P, pr_1^* DF \overset{L}{\otimes} pr_2^* F) \to H^0_P(\Delta, DF \overset{L}{\otimes} F)$$

is zero. But, using exact sequence of cohomology with support in Γ we have the commutative diagram:

$$\begin{array}{ccc}
H^{-1}(X \times X_P - \Gamma, pr_1^* DF \overset{L}{\otimes} pr_2^* F) & \to H^0_\Gamma(X \times X_P, pr_1^* DF \overset{L}{\otimes} pr_2^* F) & \to 0 \\
\downarrow & \Big\downarrow r' & \\
H^{-1}(X \times X_P - \Gamma - X_2, pr_1^* DF \overset{L}{\otimes} pr_2^* F) & & \\
\downarrow r & & \\
H^{-1}(\Delta - P, DF \overset{L}{\otimes} F) & \to H^0_P(\Delta, DF \overset{L}{\otimes} F) & \to 0.
\end{array}$$

To prove that $r' = 0$, it suffices to prove that $r = 0$.

Let us study the case: F free constant of rank one on $X_P - P$. Let us remove the shift of degree and the twist by the sheaf μ_n. Denote by $A_{X\times X_P - X_1}$ the constant sheaf $A = \mathbb{Z}/n$ on $X\times X_P - X_1$ extended by zero on $X\times X_P$. We have (non canonical) isomorphisms:

$$H^{-1}(X\times X_P - \Gamma - X_2, pr_1^* DF \overset{L}{\otimes} pr_2^* F) \tilde{\to} H^1(X\times X_P - \Gamma - X_2, A_{X\times X_P - X_1})$$

We have to prove that the restriction morphism

$$H^{-1}(\Delta - P, DF \overset{L}{\otimes} F) \tilde{\to} H^1(\Delta - P, A).$$
$$H^1(X\times X_P - \Gamma - X_2, A_{X\times X_P - X_1}) \overset{r'}{\to} H^1(\Delta - P, A)$$

is zero.

But the exact sequence of sheaves

$$0 \to A_{X\times X_P - X_1} \to A \to A_{X_1} \to 0$$

yields a commutative diagram

$$\begin{array}{ccccc} H^1(X\times X_P - X_2 - \Gamma, A_{X\times X_P - X_1}) & \to & H^1(X\times X_P - X_2 - \Gamma, A) & \overset{v}{\to} & H^1(X_1 - P, A) \\ \downarrow r' & & \downarrow u & & \\ H^1(\Delta - P, A) & \tilde{\to} & H^1(\Delta - P, A) & & \to 0 \end{array}$$

in which the first row is exact.

It is therefore sufficient to prove that $\ker(u) \supset \ker(v)$.

In terms of coverings, this can be re-interpreted in the following way:

Proposition. *Let $X' \to X\times X_P - X_2 - \Gamma$ be a Galois covering of group $\mathbb{Z}/n$ such that its restriction to $X_1 - P$ is trivial. Then its restriction to $\Delta - P$ is trivial.*

In the general case, the sheaf F on $X_P - P$ is defined by an operation of $\pi_1(X_P - P)$ on a group of the type $(\mathbb{Z}/n)^r$. By a transfer argument and an easy dévissage, M. Artin reduces the proof of proposition 6.1. to the previous case i.e. F constant over $X_P - P$ of fiber $\mathbb{Z}/l$ (l prime). However in the process of untwisting the sheaf F, he has to take ramified coverings of X_P, so that to prove the proposition 6.1. we have to prove a slightly more general proposition that the previous one:

Proposition. *Let* spec(A)/spec(k) *be the spectrum of a strictly local (i.e. henselian and the residue field k is separably closed) regular ring of dimension 2, $P \ni$* spec(A) *its closed point; X_1, X_2, X_3 subschemes defined by three regular parameters in general position (any two of which yield a basis of* *the cotangent space), and Y a closed subscheme of dimension 1 in general position with respect to* X_1 *and* X_2. *Let* $V \to \mathrm{Spec}(A) - X_1 - X_2$ *a Galois covering of group* $\mathbb{Z}/l$ *which is trivial over* $X_3 - P$. *Then V is trivial over* $Y - P$.

Let x and y the parameters of X_1 and X_2. It follows from the Purity theorem [3] that the covering V is defined by the equation:

$$T^l = x^q y^{q'} \qquad 0 \leqslant q, q' \leqslant l.$$

The covering V is trivial over $X_3 - P$, if and only if:

$$q + q' = l.$$

Let $Y' \to Y$ be the normalization of Y. The scheme Y' is strictly local regular of dimension 1. If P' is the closed point of Y', we have a commutative diagramm:

$$\begin{array}{ccc} Y' & \hookleftarrow & Y' - P' \\ \downarrow & & \downarrow r \\ Y & \hookleftarrow & Y - P \end{array}$$

The scheme Y being in general position with respect to X_1 and X_2, one sees easily that x and y induce on Y' sections which differ by a unit. Therefore the equation of the covering induced by V over $Y' - P$ is:

$$T'^l = u$$

where u is a unit on Y'. This covering can be extended to an étale covering of Y' and since Y' is henselian, this covering is trivial.

References

[1] Artin, M.: Grothendieck Topologies. Mimeographed Notes, Harvard, 1962.

[2] Hartshorne, R.: Residues and Duality. Lecture Notes in Mathematics. Berlin-Heidelberg-New York: Springer 1966.

[3] Artin, M. and A. Grothendieck: Séminaire de géometrie algébrique de l'Institut des Hautes Etudes Scientifiques (1964).